Your Mind is a FUNNY THING

WHY WE LAUGH and HOW TO BE FUNNIER

Copyright © 2014 by David Martin Gutteridge

Your Mind is a Funny Thing

First Edition

All Rights Reserved

ISBN-13: 978-1502515650
ISBN-10: 1502515652

Cover design and interior graphics:

Copyright © 2014 by David Martin Gutteridge

Cover profile photo by John Latorre

Dedicated to my Mom and Dad,

who, in their separate ways,

taught me something more important than what to know:

How to learn.

Acknowledgements

My sincere thanks to all the following people who helped me with this book.

My cousin Dr Vanessa Auld, who messes with the minds of fruit flies at the University of British Columbia, for her patient answering of all my neuroscience questions. Cees van Leeuwen, professor at KU Leuven, who was at Riken BSI at the time I spoke to him, for his explanation of how the brain works holistically. Jane Clark for her very detailed proofreading of my book. I had to make yet more changes after she was done, so any errors still present are mine, not hers. My friends Wayde Compton, Dax Oliver, and Chris Wells for their feedback on my far too lengthy early drafts. Thanks also to the Tokyo Comedy Store and all its members, for being my laboratory. I couldn't possibly list all the standups and improvisors who have helped me get funnier over the years, but to a name a few, Spring Day, Huw Lloyd, Michael Naishtut, and Ken Suzuki. For the audiobook version, Lindsay Nelson, Vinay Murthy, Jon Sabay, and Aziz Vora for reading the quotes that accompany every section. Jack Merluzzi for putting up with all my audio recording and editing questions.

Lastly, thanks to all the countless people who have had to suffer me over analyzing comedy for the seven years or more that I spent working on this book.

Table Of Contents

1. Laughter 3
- What's So Funny? 3
- Resistance Is Futile 6
- Potential Humour 8
- Schrödinger's Punchline 12
- A Joke By Any Other Name 16
- Standup Would Be So Much Easier If I Could Just Tickle You 20
- Sincerity 22
- Monkeys Who Can't Find The Remote 25

2. Funny 31
- Goldilocks And The Three Jokes 31
- The Laugh Centre 33
- Making Up Your Mind 35
- Maybe Horses Are Funny 36
- Patterns Of Behaviour 38
- One Word, Plastics 40
- Survival Of The Connected 42
- Memory 44
- Dam That's Funny 46
- The Punchline 50

3. Humour 55
- But... Why? 55
- Don't Stand So Close To Me 61
- False Positives 64
- Gotta Get Me Some Of That 69
- He Who Laughs Last 74
- The More Or Less You Know 76
- I Don't Get It 79

4. Comedy 85
- Two Types Of People Walk Into A Place And Say A Thing 85
- Back Up A Sec 86
- You Can't Make Me Laugh 91

Fuck...94
Comedic Uranium..97
Would Comedy Exist In Utopia?.................................100
Giggling At Funerals..103
Roots...105

5. Timing...111
Are We There Yet?..111
Microtiming..112
Mesotiming...114
The Myth Of The Long Set Up...................................117
Before It Gets Old...120
Macrotiming...122
Say No More..125
Timeless Jokes...128
Another Term For "Not Funny Anymore"................131

6. The Comedian..137
No Comedian Is An Island..137
What If Richard Pryor Was From The Suburbs?....140
Women..146
The Outsider..152
A Computer That Can Be Programmed To Tell Jokes.....................153
Holy Ground..155
Irony..158
Zaniness..164

7. The Audience..169
The Stepford Audience..169
Improv...172
Please, Have Another Drink.......................................175
Hecklers..179
Five Minutes Of Grace...183

8. Funnier..187
Making You Laugh... With Science!..........................187
Connoisseurs Of Comedy..191
How To Be Funny With Your Friends........................195

How To Learn..201
Finding Your Audience..203
Knowing What Works..205
It's All In The Delivery…?..209
Comedians Need You More Than You Know.................................210
Stimulus And Response..213
A Page More Blank..216

Bibliography..221
Index..227

Chapter One

LAUGHTER

2

1. Laughter

☺ What's So Funny?

"If they laugh, it's funny."

~ George Burns

Have you ever been at a restaurant when a table of people nearby are laughing hysterically at their own conversation? If you've ever listened in to find out what was so hilarious, then most likely you found that what they were saying wasn't all that funny. Sure, it might be that you don't know the story behind some inside jokes, and maybe they're a little drunk, but those excuses only go so far. Really, it just seems like what they're laughing at is stupid.

If you've ever had an experience like that, then you've partially replicated a scientific study carried out by Professor Robert Provine at the University of Maryland Baltimore County. One day, with the help of some assistants, he decided to go around to various public places to listen in on people conversing. When the research team heard someone laugh, they would write down the statement that came immediately before, the words that presumably triggered the laughter. He discovered, much like when you've been eavesdropping on others at restaurants, the statements which triggered laughter are, far more often than not, completely mundane. The kinds of phrases that on their own don't seem to indicate any great comedy at work. Things like, "I'll see you guys later," and, "Do you want one of mine?" and stuff like that. According to Provine, it's not the case that there was an obvious wider context not being described that made these statements work as comedic punchlines. Although there were some cases that the researchers could identify as attempts at comedy, Provine tells us that the assistants helping him in his study estimated that only about ten to twenty percent of prelaugh comments were even remotely humorous.

Provine concludes that, "most laughter is not in response to jokes or humor," which is in agreement with the most recent science. Laughter has a social function in helping us form and deepen bonds in our social groups. The question remains, though, as to what exactly it is we are laughing *at* in order to facilitate the social benefits of laughter. There still needs to be a reason why some, and not all, things said between friends cause laughter. On top of that, we seem to be able to discern some quality that differentiates something intended to be funny and something that accidentally ends up being laughable. Just about anyone can intuitively see that there there is a qualitative difference between what Provine describes as "humourless prelaugh comments" and the sharply crafted words of a standup comedian on stage. Does it make sense, though, to describe something as "humourless" if it gets a laugh?

A few years ago, after a comedy show I performed at, a guy who had been in the audience wrote a blog post in which he described how much he *hated* the show. We were all a bunch of amateurs, we weren't funny, we just sucked. What really stood out to me in this blog post, though, is that the author writes, and I'm quoting his words exactly, that he "couldn't understand why the other 80 people in the room were laughing." By his own account, he is far in the minority, but he's firm in his conviction that it's everyone *else* who doesn't understand funniness.

Humour can not be evaluated separately from laughter. As someone who has been performing standup for about fifteen years, and improvisational comedy, usually just called "improv", for about three decades, I can assert with confidence that it happens all the time that a comedian can try out a joke and half the audience laughs and the other half doesn't. Which side of the room gets to have the final say on whether or not it was "funny"? Does the person who is in a conversation and laughing have any more or less right to decree funniness than the person outside the conversation, eavesdropping with cold objectivity? Is it fair, or even meaningful, to say, "they were laughing, but it wasn't funny"?

If we go down the rabbit hole of separating laughter from funniness, it not only complicates the parameters of what we're trying to explain, it opens the door to some unappealing judgmentalism. To be truly objective about this, we need to accept that other people's laughter counts, even when there's nothing there for ourselves. If I hear someone else laugh when I don't, it's still *funny*, just not to me. I wouldn't even say the guy who hated my performance is wrong just because the other 80 people in the room laughed. It just wasn't for him.

The only objective measure for funniness was said best by George Burns, referenced at the top of this section. The "if they laugh, it's funny" paradigm unifies what is funny and lets us know how to identify it when we see it. When a friend laughs at something mundane another friend says, it's the same thing as when an audience laughs at a standup comedian. They both caused laughter, so they're both funny. Finding the objective factors that create a subjective result is a tricky thing, though, and will take a whole book to explain.

☺ Resistance Is Futile

> *"Humour can be dissected as a frog can, but the thing dies in the process and the innards are discouraging to any but the pure scientific mind."*
>
> ~ *E. B. White*

The words "humour" and "human" sound similar enough to suspect a connection, but actually they don't share the same etymological roots. Whereas "human" originates with the Latin *homo*, as in *homo sapien*, the word "humour" originates with the Latin *umor*, meaning "body fluid". You can still see that meaning used in modern contexts, like in *vitreous humour*, which is the name for the liquid gunk inside your eyeball. Through the middle ages they had this idea that health was determined by a balance of essential fluids in the body. By around the 1600s, the meaning of the word became associated with being in good spirits, presumably because your essential fluids were in balance. From there it came around to our modern usage.

Even though the words "human" and "humour" aren't related, everyone agrees that the quality of humour is integral to being human. Animals, even those that demonstrate a sense of playfulness, don't seem to tell jokes to each other. Humour separates us out, and if we're being honest about it, makes us feel superior. If you're a fan of science fiction, you'll note that it comes up a lot that sentient creatures from other planets are depicted as having little or no sense of humour, which, and I mean this on every level, *alienates* them. The same is true for fictional robots and real world computers, where their procedural logic seems to preclude their ability to join us in the playground of humour.

People like the idea that you can only participate in humour by possessing some undefinable and innate quality of being human. Laughter reassures us that we are more than mere biological organisms who just cycle through stimulus and response as we live out our lives. I think that some people fear humour being demystified because it takes away something precious. If we program a computer to play chess better than a human, then we can simply acknowledge that computers are faster at processing data than we are. If we can program a computer to make us laugh, then it might feel like it's us who are being reduced to mere robots with input and output mechanisms.

That's not going to happen though. Humour, while being explainable, is not going to turn into a colour by numbers exercise by understanding it. No more than learning grammar could automatically make a person into a great author, or that learning to play the guitar ensures rock star success.

Understanding the process of how humour works in the mind will help comedians and audiences alike get a sense of why a joke failed, why a joke succeeded, why the same joke might have worked for one person and not another, why some jokes are not funny now but could be funny later, and all sorts of things which can be helpful in creating comedy in a general sense. However, nothing in this book will, and nothing in any book can, tell you what content to put into a joke. The reality is that *anything can be funny* and *anything can be not funny*, so even after you understand why that is, comedy will still be a challenge.

The ability to get others to laugh is in no danger of being reduced from an art to an algorithm. Which might be an assurance to some, but really it's just an acknowledgement of the facts. Any attempt to explain how humour works would necessarily *have* to take into account the limitless manifestations of humour that we see all the time. When we look around at all the things that make people laugh, we can easily see that it has a near infinite variety.

Put another way, we *already know* that humour is inherently a human activity with unquantifiable potential variety. The task at hand is not to force the concept of humour into something more reductive, and therefore inconsistent with evidence, but to construct a paradigm that accounts for the irreducible variety we experience.

☺ Potential Humour

> *"Everything is funny, if you can laugh at it."*
> ~ Lewis Carroll

In 2007, Noam Chomsky, the renown linguist, was asked if language was a prerequisite for humour. "For humans, that does not appear to be the case," he said. "It's not clear, for example, why children who are amused by clowns or organ-grinder monkeys should be relying on language." He's right that language isn't necessary for funniness, but... organ-grinder monkeys? How old is that dude?

Our capacity for humour goes well beyond verbal jokes and sight gags. Without worrying for the moment about why, let's consider all the possibilities, leaving no room for exception. No matter how infrequent or how specific the circumstances may be, if a human could laugh at something, it has to be accounted for in any attempt to explain the mechanics of humour. To create a truly consistent model, there simply should be no such thing as an exception.

One thing verbal jokes and sight gags have in common is that they are deliberate actions on the part of someone trying to provide comedy, but of course there are also unintentional funny situations. The ways in which we laugh at someone tripping, misspeaking, getting caught off guard, or whatever else are too numerous in their variety to account for. Of all of them, though, the one type of incidental humour that seems to be the most popular is that of a man getting accidentally hit in the balls. For some reason it's even funnier if it's a baby or toddler that does it. At least according to the consensus of countless videos on YouTube, from countries and cultures all over the world. The internet has also taught us that anthropomorphizing cats is widely agreed to be funny. Not only does humour not have to be anything deliberate, it can even be sourced from an animal or inanimate object that has no capacity for participation in human relationships.

Ever laughed because you smelled the unmistakeable smell that indicates someone farted during what was otherwise supposed to be a formal event, like a business meeting or funeral? Maybe you haven't laughed at something like that since you were twelve years old, or forty something if you're me, but it still counts as laughing at an olfactory cue. I've seen musical tones make people laugh, and, when I was in art school, abstract paintings. I don't see any reason why one's own body position couldn't make someone laugh. Could a chef make a joke based on a taste? Maybe it would only be funny to other chefs, but that would still count. As far as I can see, if a human can perceive it, a human can laugh at it. The vast majority of what we laugh at comes in through usual channels, like words and images, but that seems to have more to do with the proportions of what things we pay attention to than it does with the potential for them to be funny. Kind of like how you're more likely to die by slipping in the shower than by lightning, not because showers are more dangerous than lightning, but because you encounter showers way more often than lightning.

There are also levels of perception that are not sensory but intellectual. For example, it happens all the time that someone will attempt a joke, but it will fail to make anyone laugh. In that situation, we know it was *supposed* to be a joke, but without the evidence of laughter, what is it that we are identifying?

There's also no reason you need input of any kind to laugh. You can just be sitting alone in the dark without doing anything particular, and maybe a thought strikes you as amusing. A memory, a vague notion, a fantasy, or whatever you can think about.

All of these things vary wildly in how likely and how frequently they could make humans laugh. None of them, however, compare in terms of reliability to the one thing that almost always gets a laugh, which is laughter itself. Study after study has confirmed that people are most likely to laugh when others start laughing. It's been shown that people trying to be funny in social situations laugh the most, more than the people they are trying to make laugh, and it seems that this is a sort of instinctive encouragement to get others to join in. Only professional comedians, and people following our cultural expectation of what someone with "good" delivery sounds like, make the effort to suppress laughter when trying to be funny. It would be accurate to say that the funniest thing in the world is laughter itself.

Anything can be funny, and anything can be not funny. And we can go even further than that. Anything *could have been* funny.

Every thought we have, whether brought to our attention via any of our senses, or even if it's just a thought that we evoke by ourselves, has the potential for humour. Sometimes we know it was definitely there because we laughed. Sometimes we understand it was supposed to have been there, like when a standup comedian bombs on stage. What we don't often think about, though, is when it *could* have been there. Out of all the mundane thoughts that pass through our head all the time as we live our lives, if any of them could have been funny, but just missed for some reason, would we know? Maybe just within today you've had dozens, or hundreds of thoughts that *almost* made you laugh, but didn't, and you didn't think about the absence of laughter because we don't usually think about things that don't happen. However, just because a category of thought hasn't made you laugh yet doesn't mean it doesn't have the potential.

If I try to stab you in the arm with a pin, and miss, you can still objectively identify the potential effect it might have had on you. If someone behind you tried to jab you with a pin, and missed, and you never knew, an outside observer could still objectively talk about what might have been. Truly understanding a stimulus and a response mechanism means being able to accurately identify when and why it missed. A fully consistent model for explaining humour also needs to be able to explain all the things that weren't funny not only in terms of the absence of the evidence of laughter but in what potential they had in spite of no response.

We want to be able to locate not only where there were identifiable *humorous occurrences*, indicated by the presence of laughter, or organ-grinder monkeys, but also the instances of *potential humour*, the same way we can identify a pin even if we've never seen it actually cause pain.

☺ Schrödinger's Punchline

"If you tell a joke in the forest, but nobody laughs, was it a joke?"

~ Steven Wright

Is it possible that within every punchline told, or funny image seen, or amusing instance that happened, there is some essential piece of information, some kind of meme, that forms a stimulus that we respond to? We often say that things "make" us laugh, which reveals both our experience of laughter as something involuntary, and also our expectation that when something is an involuntary reaction, there is an identifiable *external* cause. Like getting jabbed by a pin, and the feeling of pain that follows.

Stimulus and response is such a routine way of experiencing the world, it shapes our expectation of how the world *should* be experienced. We tear up when something gets in our eye, we jump when something startles us. The model of stimulus and response is so common and reliable that it's understandable that when we are presented with something like laughter, a largely involuntary response mechanism, we would try and identify just what the cause is and how it makes us respond the way we do. Even though no one can ever be sure what will make them laugh in the future, people are commonly quite sure about what was the cause of laughter in the past. The ability for most people to identify the experience that preceded the laughter leads to an understandable assumption that there must be something external that stimulates a laugh response. A distinct entity of its own, identifiably separate from the source and the recipient.

There's just one huge, monster sized problem with trying to peel off the outer layers of presentation and identifying exactly what the stimulus is *within* potential humour, which is that the response is nowhere near consistent. Unlike pins, occurrences of potential humour aren't consistent in terms of whether or not they will have *any* reaction *at all.* Maybe somewhere there are Shaolin Kung Fu masters who spend decades mastering the techniques of pain control so that they don't say "ouch!" when you jab them with a pin, but I think we can safely ignore the radical exceptions and state that the consistency of pain response is uniform among humans. Consistently uniform among just about all species of animal, in fact. If all the chairs in a comedy club were rigged to jab everyone with a pin simultaneously, there would definitely be reactions from everyone, leaving no doubt about the existence of the pins. It happens all the time, though, that a comedian will attempt a joke and some of the room will laugh and some won't. Further, not only can there be disagreement on whether or not a joke was successful, there can even be disagreement about whether or not it was, in fact, ever a joke at all. Very different from objective stimulus like pins. Even the Shaolin Monks who trained themselves to not say "ouch" when jabbed with a pin still all agree that it was definitely a pin that jabbed them.

If there were such a thing as a humorous occurrence that *always* got a laugh response, then the causes and process of humour would almost certainly already be as well understood as the relationship between pins and pain. Even if it were the case that a humorous occurrence was only funny the first time a person heard it, so long as it was *always* funny that first time with new audiences, then the study of humour would be in the low hanging fruit area of research projects. Just put someone in a brain scanner and trace the route between stimulus and response. Outside of the lab, there would be less need for comedians to innovate, as that old adage, "there are always new audiences for old jokes," could be taken to the extreme. Jokes that worked hundreds of years ago wouldn't fall out of favour as new audiences were born, and at some point enough jokes could get written to keep people laughing without the need to keep coming up with more.

Not only does potential humour not work with that level of consistency, most humour fails to appeal to a significant majority. One study, called *The Laugh Lab*, directed by Professor Richard Wiseman, had 350,000 people in 70 countries look at 40,000 different jokes to measure which was the most successful. The top joke, a variation on a joke that seems to have originally been written by Spike Milligan, was rated by 55% of the people surveyed as being funny.

> *Two hunters are out in the woods when one of them collapses. He doesn't seem to be breathing and his eyes are glazed. The other guy whips out his phone and calls the emergency services.*
> *He gasps, "My friend is dead! What can I do?"*
> *The operator says, "Calm down. I can help. First, let's make sure he's dead."*
> *There is silence, then a shot is heard. Back on the phone, the guy says, "OK, now what?"*

The slight majority that rated this joke as funny didn't rate it as the most funny, it was just the joke that was most commonly rated as at least somewhat funny. Professor Wiseman described the impact of the joke by saying, "Perhaps we uncovered the world's blandest joke - the gag that makes everyone smile but very few laugh out loud."

Still, the fact that a little over half of the people surveyed found it amusing does indicate that it could be possible for some potential humour to have fairly wide appeal. Also, as our world becomes more globalized and there is more opportunity for our sensibilities to converge, we could see more consensus on what is funny. However, we're still a long way off from that, and the current evidence is that while the world's top comedians find audiences big enough to make fortunes from, in terms relative to society at large, their appeal is very limited. Most humour falls very short of appealing to 55% of the available audience. As Professor Wiseman says, "If our research into humour tells us anything, it is that people find different things funny... There is no one joke that will make everyone guffaw."

Pins are so reliable as a stimulus, that we can just look at one and know what it will do in advance of using it. It seems potential humour might be similarly identifiable because we can write jokes down, describe what happened yesterday that was funny, or record a funny sketch on video. Despite all the forms of delivery, though, potential humour will not be known for sure to get the right response until after it's tried and at least somebody laughs. If no one laughs at something deliberately constructed to be a joke, is it still a joke? Was it ever? And then, if we try it on someone else, and that new person laughs, is it *now* a joke? Is the fault with the joke or the audience when it doesn't work? If I knew it was *supposed* to be a joke in spite of not laughing, does that count?

The cause of laughter is hard to pin down, pun intended. How can it be that we have such a specific response, laughter, to a stimulus that is so ethereal that we can *only* ever truly be sure it existed after it has given us results?

☺ A Joke By Any Other Name

> *"Laughter is the shortest distance between two people."*
>
> ~ *Victor Borge*

Imagine you are given a rose by someone you love. Even if you aren't the type who particularly likes flowers, I would imagine that you'll still appreciate the gesture and be somewhat happy about it. If it was someone you really liked and hoped they had the same feelings for you, the moment of getting a rose from them could be a thrilling experience. However, now imagine that the person who hands you a rose is a crazy stalker who has been waiting outside your house every night for the last two weeks, trying to catch a glimpse of you through your bedroom window. Now being handed a rose is far from a happy experience. You might even be terrified about the implied message of the rose if you thought this person was genuinely dangerous. In the two cases, your feelings are wildly different, but the rose isn't. The rose is the exact same plant in both scenarios, but what has changed is that both the person delivering it and the person receiving it are creating different contexts that alter the rose's meaning. That's why you can listen to people at the next table over and not be affected by their jokes, because it's a little like watching two lovers you don't know at all exchange gifts. The gifts aren't for you, so who cares?

Trying to define the characteristics that objectively identify love by measuring the shape and colour of roses would miss the point entirely. A rose is just a plant until someone puts it into a context of love. Similarly, potential humour is nothing more than words or images or events until someone receives it in a context that makes it funny for them.

The craft of comedy seems to operate a little differently than humour we see between people. And it is different, but only in the way that all arts have differences from the core feelings they appeal to. Like love songs, where the singer and the person listening aren't in love with each other, and, via modern media, won't ever meet or know each other. Love songs are at play with feelings of love that originally existed to bind us together in relationships, but can now be used to inspire similar feelings for our comfort and amusement. Ditto for horror movies playing with our sense of fear, which originally kept us from danger. And of course, porn, which can be very different from our real life experience of sex, but is no less based on the feelings of lust that originally existed to drive us to reproduce. Humour also had an original purpose, and we need to be careful to not confuse what we do with comedy with why we originally laughed. Just like how the fact that humans now frequently watch sex as a form of entertainment doesn't mean sex developed in order to be watched.

Comedy and humour are actually separate things. Humour is a sensation in your mind, and comedy is a craft. Humour is a fundamental human trait that everyone participates in and has an ability for. Not only can everybody laugh at anything, but just about anybody can inspire laughter in friends and family as a matter of course in their everyday relationships. Comedy is the elevation of that interaction into a performance. Just like a love song is the distillation of common human feelings into a particular shape for the sake of art and entertainment.

You would be right to point out that at least one difference between humour and love is that there is no consistent and involuntary reaction that humans have when presented with some symbolic gift of a loving relationship, whatever that signal may be. There is no equivalent in love to the laughter response. However, I don't think anyone could sensibly deny that there are strong and distinct feelings inspired by loving acts. The real difference is that humour, because of its laughter response, is easier to identify for an outside observer.

There is a spectrum of stimulus-and-response mechanisms. On one end, in the domain of pain and pins, we can objectively measure both the stimulus *and* the response, equally objectively. On the other end, where we find love and roses, both stimuli and responses are wildly variable. Humour is in the middle of that spectrum, having a response that we can hold constant, but stimuli that vary wildly. In any case, all we need to achieve with a comparison to love is that it is plausible for one human to inspire a sensation in another human based not on the particulars of their actions, but using the context of their relationship to shape the meaning of what they do. Potential humour becomes actual humour because of what we put into it and what we get out of it, not what it happens to be made of. Without the right relationship, timing, and context, verbal jokes are just words, sight gags are just things that happen, and amusing events are just sequences of actions.

The contentious part of accepting that humour is similar to the infinitely variable signals sent between lovers is that it means accepting that every performing comedian has a relationship with their audience, which seems odd given that we assume a comedy performance is usually for people the comedian does not know personally. In his book on laughter, Provine described standup as "socially impoverished". But a relationship with the audience is exactly what is going on. That an audience for a standup comedian is largely made up of strangers the comedian has never met before is a surface level description that is misleading. Even strangers bring commonalities into the start of a relationship, right from before any interaction has taken place. There are cultural conventions, environmental conditions, and a lot of human nature that act as a base for us to establish communication. A large part of the craft of the comedian is having a feel for what connections can and do exist between groups of people beyond the people they know personally. This is the dividing line between *friend-funny* and *stage-funny*, an important distinction that any experienced comedian understands very well. The reality is that a comedian *is* in a relationship with the audience, but that relationship is not with individuals *in* the audience. It's the collective entity that ensues as a result of the community that we label as an audience. It's still the same process, where humour is the result of a relationship, it's just that to elevate humour to the craft of comedy, it's the type of relationship that changes.

The form and content of things that are funny is incidental to the underlying relationship that appreciates them. Funny things can, and do, take any shape or form, and because of their lack of consistency, do not in themselves reveal the nature of humour. When we want to find out what's funny, we want to find out what is going on inside people, and between them, that makes them respond with laughter.

☺ Standup Would Be So Much Easier If I Could Just Tickle You

> *"If you prick us do we not bleed? If you tickle us do we not laugh?"*
>
> ~ *William Shakespeare*

One thing that *isn't* funny is tickling.

Wait... what? Tickling obviously makes people laugh, doesn't it? How can I say it's not funny? Aren't I now suddenly breaking my own, "if they laugh, it's funny," rule?

When I say tickling isn't funny, what I mean is that there is no humour being processed in the brain. You see, unlike when someone gets a joke, we can jam someone in a brain scanner and trace the route that tickling goes through from where they're touched to where their brain kick starts the laugh response, about as well as we can trace a pin prick. The path starts somewhere in the *somatosensory cortex*, which registers the touch sensation, and ends up in the *motor cortex*, responsible for all body movements, which controls all the vocalizations and actions that combine to form laughter. This is different from understanding and responding to jokes. If I put you in a brain scanner and got you to laugh at jokes, the pathways through your brain are much less distinct.

Scientists have tried to find those pathways, to see how the experience of getting a joke winds it way through the brain, by throwing people into brain scanners and getting them to laugh. Some of the areas within the brain that have been linked to humour are the *nucleus accumbens*, the *medial ventral prefrontal cortex*, the *bilateral posterior temporal lobes*, and many others, with equally crazy names. What's notable about them is only that these attempts to find a part of the brain identifiable as a "humour centre" hasn't produced any consensus. As one researcher, Joseph Moran, said in a study called *Neural correlates of humor detection and appreciation*, "there's no standard set of regions that we know are involved in humour."

The hard evidence of science shows that the brain processes tickling and humour differently. However, they both end up at the same place, laughter, so there has to be at least some relationship between them.

Research has shown that tickling seems to exist in other mammals, like chimpanzees, dogs, and rats. Rats, for example, have been shown to respond to tickling-like touching with a high pitched sound. It's often too high for humans to hear, so it's gone largely unnoticed for a long time. Not to mention that it's not most people's first impulse when they see a rat to tickle it. I think that the purring of domestic cats might also be included. Even though we might not ordinarily categorize scratching behind their ears as "tickling", and purring doesn't sound like laughter, it's still a sound that is usually elicited by touching. The specifics will naturally vary among species, but the implication remains that the origins of tickling might go back to a common mammalian ancestor. If so, then tickling predated humour, predated language and culture, and even predated us as a species. On the other hand, while we potentially share tickling-like behaviours with other species, humour seems to be exclusive to us. From that chronology, it seems reasonable to suggest that humour evolved within us by repurposing the available mechanisms used in tickling. Just as it's likely that crying to get dust out of our eyes was present in us as a species when we diverged from other simians, and then later it took on the extra purpose of expressing sadness.

Laughing from being tickled and laughing at something humorous are two identifiably different things. Disagree with me and I'll throw you into a brain scanner and tickle you until you cry. Later we'll reconnect tickling with humour by getting more in depth about our evolution. For now, though, to understand why we laugh and what is humorous, we need to set tickling aside, just as we need to do with other conditional factors, like the amount of sincerity in the laughter.

☺ Sincerity

> *"The secret of success is sincerity. Once you can fake that you've got it made."*
>
> ~ Jean Giraudoux

If you've watched the TV show *The Simpsons*, then you are probably familiar with the trademark laughter of the schoolyard bully character Nelson Muntz. The defining feature of his laughter is its deliberate sing-song intonation, which unmistakably conveys a social message of putting down the target, as opposed to sincere laughter from finding something funny. In one episode, the character Lisa attempts to perform standup comedy in her school cafeteria. Her jokes are lame, so Nelson wants to laugh at her failure. The problem, though, is that using laughter to express derision to a standup runs the risk of being mistaken for praise, even with his trademark intonation. So Nelson clarifies, "The following *ha ha* is not from amusement, but an expression of contempt. *Ha ha!*"

In the real world, no one is quite as straightforward as that, so laughter used to deliberately assert status and belittle someone can often sound just like actual laughter motivated by amusement. Despite the camouflage, though, we need to acknowledge that they are very different things. Just as we put tickling to one side, we also need to separate out instances of laughter that are not sincere responses. Any human expression or response can be manipulated to fit our social goals, for good or for bad. The whole art of acting is based on the simple premise that an actor can deliberately emulate any kind of human sentiment in order to get a response out of an audience. The only difference between acting and many other social manipulations is that actors are up front in asking you to suspend your disbelief.

Consider how soccer players roll around in exaggerated pain to get a response from the referee. They may be experiencing at least some genuine pain, if the act is preceded by an actual kick in the shins, but they exaggerate it in hopes of eliciting a favourable call against the other team. Even though the genuine pain and the deliberate exaggeration are bound together, we can still identify them as two separate things. There's no way to measure how much is pain and how much is exaggeration, but we know they're both distinct factors.

Some people can cry on demand in an effort to try and gain sympathy from others, twisting what is ordinarily an involuntary response to emotional stimulus into a deliberate behaviour in order to make a situation go the way they want. Similarly, people routinely use laughter as a tool for manipulation as much as every other feeling. For example, laughter can be used to defuse a tense situation, to convey confidence, to suck up to a superior, to flirt, and many other situations. Consider how the tactic of passive aggressiveness turns conveying gratitude or acceptance completely upside down by using overly deliberate agreeable behaviour to convey displeasure or anger. There is no limit to the degree which humans can and will bend and manipulate any and all methods of expression to further their personal goals, and laughter is no exception to that. In any situation when a human can gain something from manipulating a natural response, would we ever be able to objectively draw a hard line between the sincerity and the deliberation? Situations of bullying, flirting, defusing tension, and whatever else are the results of circumstances of infinite variety, with uncountable devils at work in the details.

In one instance I can laugh completely sincerely at a joke by someone with no motivation other than thinking it was funny. In another instance, I can completely force a laugh at a joke told by somebody I'm trying to get into bed with, because I'm a horrible person. Somewhere in the middle is yet another situation where I want to sleep with the person speaking, but I also genuinely found what they say at least a little funny. The laugh contains some amount of genuine amusement mixed with some amount of manipulation. In the real world it's often impossible to clearly separate the two motivations, but they can still be said to be two distinct things.

Just as comedy is the elevation of humour into a craft, we can view bullying, tension alleviation, sucking up to superiors, and any other deliberate manipulation of laughter as repurposing humour for social needs. It's like comedy gone bad, using humour as a tool to gain an advantage instead of trying to entertain. From that point of view, how laughter relates to these social manipulations would be a worthwhile study, but that's not what this book is about. This book only seeks to explain the *sincere* act of finding something funny, to the degree that there's no additional social objectives involved. I'll leave it to someone else to write *Your Mind Is A Manipulative Asshole*.

☺ Monkeys Who Can't Find The Remote

> *"Fifty-seven years in this business, you learn a few things. You know what words are funny and which words are not funny... Words with "k" in them are funny... Cupcake is funny. Tomato is not funny. Cookie is funny..."*
>
> ~ Willy, from Neil Simon's The Sunshine Boys

The idea that words with a K in them are funny floats around the comedy world, and was even tested a little by Richard Wiseman when he went looking for the world's funniest joke in the *Laugh Lab* experiment referenced earlier. People who were shown this joke:

> *There were two ducks on a pond. One said, "Quack," and the other said, "I was going to say that."*

... generally responded better than this version:

> *There were two cows in a field. One said, "Moo," and the other said, "I was going to say that."*

Are hard consonant sounds, or "plosives" as linguists like to call them, funnier than other sounds? Is "quack" funnier than "moo"? Maybe birds are just funnier to talk about than mammals.

It could be true that certain sounds can make a phrase a little funnier, but of course no one really thinks humour is just a matter of saying as many hard consonants as possible. There are more reasonable attempts to explain what's funny, and lots of them. It's often suggested that humour is all about status. Another common notion is that humour is a matter of surprise. You've probably heard people say that what's true is funny. Many assume humour is differentiated by, and therefore based on, culture. A lot of science is predicated on the idea that humour is linked to language. Explanations of humour go back at least as far as Aristotle, who seems to have believed humour was laughing at the misfortune of others. More current thinking is that comedy is all about "benign violation", which is essentially making dangerous ideas safe. There are formal ideas, such as "incongruity resolution", the idea that it's all about breaking expected patterns. And there are intuitive ideas, like how everybody knows comedy is all about timing, even though we might not know exactly how or why.

All of these theories have a hard time covering all the possibilities. Status can't explain why puns and silly word play are funny. Language doesn't adequately cover why people laugh when a dude gets kicked in the balls. Nor does culture, because as far as I can tell, getting kicked in the balls is laughed at everywhere. There's also the problem of adequately accounting for non-funny situations, such as why people can be surprised without laughing, or why resolving an incongruity is more often just resolving a problem and not creating a joke.

I could go on, and go into more detail, but it's sufficient to say that we don't already have a reliably working model of how humour works that everyone agrees on. Significantly, all these explanations only apply in retrospect, they have no predictive ability. No one has been able to use any existing model of humour to create a formula for making new jokes. Of course, no formula can ever exist, just like there is no sure fire gift you can give someone to make them love you. This highlights the problem with just about all of these explanations, which is that they attempt to explain humour by explaining the content of jokes. Like explaining love by measuring flowers.

Roses become symbols of love when passed between the right people, but so could carnations, or lilies, or any other flower, or anything else. Similarly, humour can use status, surprise, culture, language, or anything else as a method of delivery. Some of these contexts are so frequently used for humour that one could be forgiven for mistaking them for the whole thing. Status, for example, is something that we're so overwhelmingly concerned with as a species in almost all aspects of our lives that it shouldn't be too much of a surprise that we see it exploited for humour so much.

Explaining humour by trying to find commonalities, such as status or "benign violation", in among all the wildly variable things people will laugh at, has the same problem as if people tried to understand how a television set works by seeing what is common about all the shows that appear on it. Imagine what crazy notions people might have about what was inside a television if they took the images on the screen as being somehow representative of what went on behind the screen. Starting with television shows as a way of working toward an understanding of television sets seems pretty silly only because we built televisions ourselves and so we can be smug about the distinctions between medium and content. We didn't build our brains, though, so it's harder to be sure about the dividing lines between perceptions and processes. Since we figure out a lot of what goes on inside a brain by looking at how people behave, when it comes to understanding our minds, we're actually a lot more like chimpanzees watching television than maybe we'd like to admit.

What we want to do is find an explanation for comedy that is designed from the start around the observable evidence that *anything can be funny* and *anything can be not funny*, and the only way to know when we have funniness is when we have sincere laughter. If we do that, then we can encompass all the existing notions about what humour is, and more. The place where all that can happen is the brain. It's the television set that displays all our shows about status, surprise, language, visual gags, culture, incongruity, and any other format humour can be delivered in.

We can do better than chimpanzees dazzled by all the activity on a screen. We can open up the back of the television set and see what's really going on.

Chapter Two

FUNNY

30

2. Funny

☺ Goldilocks And The Three Jokes

"You want to know what's not funny? Thinking about it."
~ Chris Rock

Jerry Seinfeld once described telling a joke as setting up a gap between two conceptual cliffs, and the audience has to make a jump over that gap. If the distance is too short, then it's too easy, so there's no challenge or excitement. It's like a joke that is boring or obvious. If the gap is too far, then you can't reach the other side. Like a joke that's too weird or unfamiliar, so the audience is just left shrugging. But if the gap is just right, so that when the audience makes the metaphorical jump, they *just barely* make it to the other side. It was an exciting jump that was fun to do. Getting a joke is a similar matter of it not being too "close" that it's obvious, and not too "far" as to be beyond the ability of the listener to get it. It's not a complicated, nor very controversial model for evaluating potential funniness.

Turns out that this analogy is extensible beyond the finely honed verbal jokes Seinfeld uses in his act. In any situation, in any media, in any interaction, in any relationship, anything that you can perceive through any of your senses, anything that is potentially humorous follows the same model. Any thought that is either too obvious or unfamiliar, *too close* or *too far*, does not inspire laughter. Most of our day to day thoughts fall into these categories, but concepts that are just right, wherever they come from, if they land on the razor's edge between the two possibilities, can inspire laughter.

Seinfeld's gap model works as a handy tip for aspiring comedians to think about how to construct a joke, but to go beyond that, to give it any credibility as a foundation for humour in a broader sense, it needs something more concrete than just a metaphor. There is a physical process in the brain that makes Seinfeld's description manifest, and this chapter will outline exactly *how* it works. Then, armed with that knowledge, the chapter following will explore *why* this humour process matters to us.

Before we explore the inner workings of the mind, though, there is just one caveat, which is that unfortunately, current technology does not provide us with the ability to see deep enough into the brain to see the process of humour in action. Even when you've heard of particular brain parts being linked to humour, as in some of the studies I referenced earlier, they're only trying to describe where things might be happening, without any ability to see how. The best brain scanners science has can see down to a resolution of about one cubic millimetre, which may contain tens of thousands of brain cells. The activity that we'll be talking about is most likely much, much finer than that, occurring on a level *between* brain cells.

Why, then, go into such detail, pushing a little past the boundaries of measurement? Partly because doing so will set up a vocabulary that can be used to make sense of a lot of the topics within the book. Much more importantly, though, the key point is to see exactly how a sense of humour could work on a *completely* biological level. If we can, then that completely divorces us from the need to explain why the content of any one joke is funny. That puts us in the position of looking inside the television set, and not at the flickering screen.

☺ The Laugh Centre

> *"Does the brain control you or are you controlling the brain?"*
>
> ~ *Karl Pilkington*

Stephen Colbert, on his television show *The Colbert Report*, once asked Steven Pinker, a cognitive scientist who writes lots of popular books about the mind, to summarize how the brain works in five words. "Brain cells fire in patterns," was Dr Pinker's response. It'd be great if we could just leave it at that, but to see where the humour is happening, we'll need to know a little more than just five words.

At least one part of how humour works in the brain, the laughter part, is reasonably well understood. Laughter is a physiological response driven by identifiable modules in the brain that can be observed using a brain scanner. Laughter itself is a cluster of different actions, including most obviously the spontaneous vocalizations we know and love, but also other associated actions like spasms, the release of pleasurable chemicals in the brain, and an increased heart rate. All of which will vary in intensity between individuals, but are consistent enough, especially the vocal component, that there aren't really any big surprises about what laughter is and where it happens in the brain. By and large, the parts of your brain that regulate those laughter outputs are within the *motor cortex*, which is the section of your brain that controls all of your movements. Whether you consciously decide to move your hand for a deliberate purpose, or unconsciously and reflexively pull it back from a pin prick, the signal to move your hand will go through the motor cortex. Similarly, when you are laughing, each action that combines to form laughter is being driven by various subsections within the motor cortex. For the purpose of understanding humour, it's good enough to think of all the associated modules as being one thing, a "laugh centre". In the process of humour, by the time the activity in your brain has reached the motor cortex, your brain is just generating output.

Finding where the humour happens in the brain *before* the laughter is the tricky part, because unlike tickling, it's hard to create a consistent input so that you know where to start tracking the sequence of activity. Sure, I can tell you some verbal jokes or maybe show you funny pictures while you're strapped into a high tech scanner. But how can I reliably reproduce for you the experience of how funny it was that time your friend accidentally revealed an embarrassing secret? Or when you laughed spontaneously because of a memory unprompted by anything you were particularly aware of? Or countless variations of "you had to be there" moments? Lab conditions restrict the kinds of potential humour we can test, leading to results that very likely say less about humour and more about the constraints of the experiments.

Even within those narrow confines of testing only certain types of jokes in labs, different studies offer different proposals of where humour happens, so we still don't see any one part of the brain light up consistently enough to be a potential "humour centre". That being the case, I think we're justified in wondering if there's another approach we can take to search for humour that doesn't rely on a modular model of the brain. Fortunately, there are other ways of looking at how the mind operates that are more consistent with the idea that anything we can think of in almost any part of our brain can be funny. Humour isn't just some switch somewhere in our brain that is simply toggled on or off with the right inputs, it's our whole mind that has a sense of humour.

☺ Making Up Your Mind

"As soon as you have made a thought, laugh at it."
~ Lao Tsu

Not all thoughts are processed the same way. A pin prick will stimulate nerves that send signals through largely predetermined pathways in your brain that are hard wired. This kind of rigidly structured process has advantages, because if someone was trying to stab you, it's good to have an automatic response that makes you jump out of the way as soon as possible. If you had to stop and ponder the implications of a knife attack, you would almost certainly end up bleeding more than you could survive before the right decisions got made. Other thoughts, like comparing which mobile phone service plan is ripping you off the least, don't have set parts of the brain with predetermined functions. Which is also advantageous, in that it would be impossible for our brains to have an automated response to every possible scenario in the universe that is not yet known. Our brain's ability to carefully consider the implications of an infinite variety of input makes us very flexible and adaptable. Having both forms of thought, rigidly automatic and flexibly processed, is a great advantage in that we can usually apply the most appropriate type to the situations they're suited for. Where things get tricky, though, is that there isn't always a hard boundary between the two.

The presumption of a lot humour research is that humour must fall into the first category of brain response, the type with at least some part dedicated to listening for some particular input, because a sincere laugh response seems to be involuntary, which is similar to other automated responses. However, as described in detail in the previous chapter, potential humour is neither as consistently identifiable as other stimuli, nor has any one particular route for receiving humorous input in the brain been identified using brain scanners. Which indicates that maybe humour is more like the adaptable and holistic type of brain activity. That doesn't seem quite right either, though, because most types of complex thinking don't usually have a specific response similar to laughter, just as there is no single response for loving someone.

Maybe, then, it's a little of both. As mentioned, humour sits in the middle of a gradient of response systems, between pin pricks and the feeling of love, in that it has a specific response but no one specific stimulus. It's conceivable that this dual nature of humour, being general in input but specific in output, could be reflected in how the brain physically processes potential humour.

☺ Maybe Horses Are Funny

> *"Humor is not a mood but a way of looking at the world."*
>
> *~ Ludwig Wittgenstein*

In 1998, while some doctors were doing some neurosurgery on a 16 year old girl to help her with epileptic seizures, they seemed to discover a "laugh button" on the front left side of the brain. When they stimulated this one area, the young girl would laugh. Testing further, "Whatever she was doing at the time she would attribute the laughter to that activity," said Dr Itzhak Fried, the lead doctor in the operation. "If she was looking at a picture of a horse, she would say the horse was so funny."

It's tempting to jump on this as being evidence that there is a specific humour section of the brain, and it's right there where Dr Fried was poking it. However, we already know from lots of other research under brain scanners that no single place consistently correlates with humour processing. If this spot that Dr Fried was poking was *the* humour centre, then surely it would have been a reliable feature in all the brain scanning that's been done on people laughing. Since it hasn't, though, we have to wonder, is there another way of thinking about this? One where we can account for no one specific part of the brain being being responsible for processing humour, yet at the same time be stimulated by poking a particular spot, as in the case of this one girl.

Dr Fried provides just such an alternative possibility by suggesting, "we just tapped into perhaps one area in a very complex network." That network concept is key in linking the holistic to the specific. Instead of one elusive module somewhere, there is a humour network that is distributed all over the brain, and these doctors were rattling one part of it. Like tapping one edge of a spider web, which sends vibrations throughout the whole thing, alerting the spider to the presence of something in need of killing and eating. The web is the humour network and the spider is the laugh response.

This gives us the model we need to satisfy the dual nature of humour in that it has a general and non specific input with one specific output. The humour network weaves throughout the brain, and because of its distributed and pervasive nature, it can be involved in just about any kind of thought. That allows for humour to have a relationship with everything the human mind can think of, which is anything. On the other end, all roads lead to Rome, in that the network ultimately connects down to some place in the motor cortex, where it triggers the response that is the same among all humans, laughter.

Now we know that a laugh response can be triggered more or less anywhere in the brain, but what is this network made of, and what *exactly* triggers it?

☺ Patterns Of Behaviour

"I like nonsense; it wakes up the brain cells."

~ Dr Seuss

The humour network is made up of *neurons*, the cells that are most responsible for all the thinking that goes on in your mind. Different neurons have different dedicated tasks, and the neurons in your humour network are listening for a specific kind of activity that exists between other neurons. To understand what that activity is and why it is meaningful, we'll need to drill down just a little further and see how neurons work, and how they talk to each other.

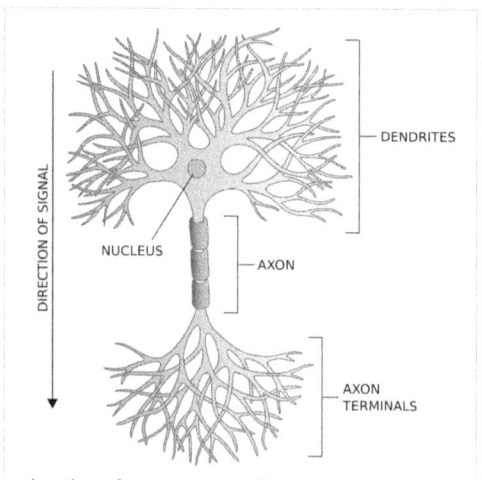

A simple neuron. Real neurons are usually much longer and have thousands of branches, and are tangled up with thousands of other neurons.

The easiest way to envision a neuron is that it's shaped something like a tree. At the top are branches, called *dendrites.* The trunk of the tree is called the *axon.* Lower down, the axon splits and branches out into many roots. The tip of each root is called an *axon terminal.* Much like a tree, the dendrites and the branches of the axon divide and multiply so they can have thousands of tips. Unlike trees, though, neurons have all these branches and tips for the purpose of being in contact with each other, in order to communicate. Signals go one way with neurons, in through anywhere on the surface of the cell, and out through the axon terminals, which are touching other neurons. The point at which axon terminals touch other neurons and send signals to them is called a *synapse,* because scientists can't ever just let things have a regular name, like "connection", which is all a synapse is.

Synapses can form anywhere on a neuron's surface, so your brain cells are all squished up together in a big wet and stringy tangled mess. A mess that is constantly tangling itself more and more, because, like all cells in your body, neurons are living entities in their own right, so they're *constantly moving.* They don't travel from place to place like blood cells do, but their roots and branches, the dendrites and axon terminals, are always groping around in the dark, sliding along the surfaces of nearby neurons looking to make new connections, and reaching beyond their current neighbours in an effort to find new neuron partners.

Neurons, like people, like having lots of friends, and are always looking for more. Also, like people, when neurons get gossip from one friend, they love to tell it to everyone they're in touch with. For neurons, gossip comes in the form of electrical and chemical signals transmitted at their synapses. When a neuron receives an electrochemical signal from any one source, it gets all excited, and it will respond by making its own signal to broadcast to all the other neurons immediately connected to it. In this way, signals flow through the brain, being passed from neuron to neuron, so that at any one time a certain number of cells will be "lit up" with activity. If you used a fancy brain scanner to take a picture, the different degrees to which different areas in your brain were lit up or not form an overall pattern representing your state of mind in that one moment. Just like one still image from a movie wouldn't be enough to know the story, though, it's how that pattern changes and moves over time that really matters. The flow of this pattern over time is the language of your thoughts.

☺ One Word, Plastics

> *"Neurons that fire together wire together."*
> ~ Donald Hebb

You probably have some people you like to talk to all the time that you want to have easier access to, and other people you're less enthused about that you don't care to contact so often. In my phone, it automatically keeps a log of the people I've spoken to recently, and I can use that list to call them again more easily the next time. It's one way that the people I talk to more frequently are easier to connect with. Your neurons are doing the same kind of thing. After all, if you had to think about something once, you might have to think about it again. If you think about it twice, then you're even more likely to have to deal with it in the future. If your brain can arrange it so that the next time you think about something the activity will flow through faster and easier, then you'll be able to respond quicker, and with more nuance.

The way neurons make closer connections to their best friends is by making more synapses at a point of connection that is getting a lot of activity. If a synapse gets lit up with activity over and over, the neurons involved will respond by adding more synaptic connections to that same spot. More connections is like having higher bandwidth for your internet connection, it allows more data to flow through. The same principle also holds true in the opposite direction. If a connection doesn't get used enough, the neurons involved will start retracting connections, and can eventually abandon it entirely. Use it or lose it. Losing connections is a good thing, because those loose axon terminals become available for use in other synapses, allowing you to learn new things in place of ideas that don't matter.

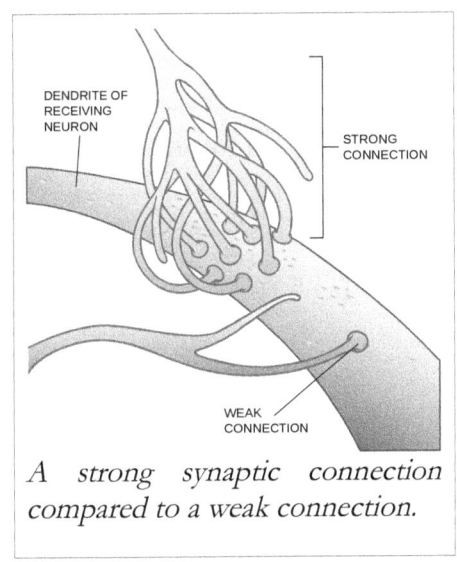

A strong synaptic connection compared to a weak connection.

This constant moving around, making and breaking connections, is referred to as *neuroplasticity*, the capacity for your mind to change. It's because of neuroplasticity that your brain has the ability to learn and improve and get better at things. And laugh at them.

☺ Survival Of The Connected

> *"Brain: An apparatus with which we think we think."*
> ~ *Ambrose Bierce*

Neurons have no idea in advance which other neurons they need to touch in order to complete networks that could be useful to you. Neurons have no thoughts of their own, and no concept of the overall mind that they facilitate. They're just blindly growing and moving, groping around in the dark, making connections as best they can, only because they like getting electrochemical hits from their friends. That your sense of self arises from all that activity is just a happy result for you, but nothing that your neurons are bothered about.

Current estimates are that within your brain, about a million new synaptic connections are made every second. Presumably, a similar number of synaptic connections are abandoned as well. Since a neuron doesn't know when it forms a new connection whether or not that connection will be a keeper, then to a certain degree, the process of finding new connections is random and arbitrary. The majority of attempted connections probably aren't very useful and get abandoned again pretty soon, but by testing out a million new connections *every second*, the chances of finding some good ones go up.

Some researchers even suspect this process of filtering the useful from the random is the foundation of even our basic sensory input. The theory is that when babies are born, they experience *neonatal synethsesia*, which is an inability to make even the simplest distinctions in perception, like the difference between sight and sound, or how to move one's limbs. As babies vocalize, feel around, and look at things, they start to sift through those connections and build more efficient pathways for the activity in their brain. When you think about how a baby's brain has no idea how to walk, and compare that to how an adult can run down stairs while talking on their phone and without spilling their coffee and not even consciously think about the motion of their feet, you can see how the selection of useful connections from random connections takes some time, but ultimately leads to some very sophisticated processing abilities.

When you conceptualize something in a way you haven't done before, somewhere in your brain are some synapses that were weakly connected and are now being lit up as part of a new pattern of activity. Note, though, that a new concept doesn't necessarily mean learning a whole new fact, it could just be a very subtle way of considering a known fact, or even just a new feeling about something well known. The process of developing neuronal networks drives all your ideas and feelings and perceptions, without regard for the kind of thought it is.

At any one time, you have countless synaptic connections on the edge of being either abandoned or strengthened, depending on whether a new pattern of thought comes along to make use of them. It's the availability of all the weak connections in your brain that's going to have an influence on how hard you laugh at something.

☺ Memory

"Humor, to be comprehensible to anybody, must be built upon a foundation with which he is familiar."

~ Mark Twain

The activity in your brain and the structure it flows though influence each other. As activity determines which connections stay and go, over time the structure of the neuronal network is altered. The shape of the neuronal network then influences the direction of the flow of activity in the future. In this way, your current thoughts help determine who you become, and who you are helps determine how you think, in a constant back and forth that dynamically forms your identity.

The flow of electrochemical activity could be said to represent your current thoughts, right now, in this moment. The structure of the neuronal network could be said to represent your personality and how you as an individual interpret your current thoughts. A bit of a simplification, but true enough, and the model that will be used going forward. The brain's structure is you, the activity is your thoughts, and the neuroplastic nature of making and breaking synapses is the cutting edge where thoughts will either disappear as fleeting notions or become a part of you.

Your personality, your memories, and the context in which you receive input from around you is dependant on your particular brain structure. Where you have a flow that goes through well established channels, with strong synaptic connections, those thoughts, in essence, are obvious to you. Where you don't have channels at all, the flow of thoughts will run into dead ends, leaving you unable to follow. In other words, ideas that are too weird. Where you have a lot of weak connections being lit up by activity, you have a thought that was neither too obvious or too weird.

Seinfeld's model of a gap is now physically manifest in the brain. Jumping across that metaphorical gap and just making it is a matter of getting connections that are, in essence, only just able to make it from neuron to neuron, by passing through synaptic connections neither too well established nor disconnected.

That flow doesn't have to be any particular *type* of thought, though, and that's how we go beyond a model for jokes and reach a model for anything that has potential humour. The flow of activity in your head can represent any kind of thought, feeling, perception, or concept. Which means that *anything* can be funny.

When you think of how it is that on a cellular level the construction of our brains are completely unique and changing every moment, it's kind of amazing that any two humans can look at the world the same way and come to a consensus on *anything*. However, not only is it possible, our brains have evolved to do just that so that we could cooperate in groups and get to where we are now. Without that capacity for coming to a consensus on things, we wouldn't have accomplished more than any other species of animal on the planet. Nonetheless, it's no wonder that two people can respond entirely differently to the exact same potential humour, which explains why people don't respond to humour with the same consistency that they react to pins. Even when two humans do laugh at the same thing, we can tell that the flow in their minds had a similar quality in terms of strength of signal, as indicated by the laughter, but the specific neuronal structures that underlie the flow of mental activity are almost certainly very different. The craft of the comedian, to get whole rooms to come to a consensus and laugh, is no small task.

☺ Dam That's Funny

> *"It is easy to forget that the most important aspect of comedy, after all, its great saving grace, is its ambiguity. You can simultaneously laugh at a situation, and take it seriously."*
>
> ~ Stephen Fry

How exactly could a humour network determine if the activity flowing around the pathways of your brain had just been through weak connections or strong connections? In the case of potential humour being *too far*, it's easy to see how the humour network would *not* pick up any activity worth noting. *Too far* means the signal heads down to dead ends where there are no connections, and so the signal stops before the humour network has a chance to evaluate it. In the other two cases, though, *too close* or "just right", there has to be some kind of qualitative difference between a signal that had just passed through a set of weak synaptic connections, and a signal that had flowed through strong ones.

Imagine a flowing river, with a dam at one end of it. The dam has a gateway on it to allow some of the water to pass through. Open the gateway wide, and a powerful stream of water comes blasting through. Close the gateway down to the point where it's only open a little, and only a small trickle of water passes through. If you put some kind of sensor on the side of the dam where the water comes out, you could determine the degree to which the gate was open from the speed and pressure of the water coming out of it. Similarly, as signals travel around the brain they can be strong or weak, depending on how many neurons are involved. However, just knowing the strength of a signal is not enough. How would the humour network differentiate between a strong signal that had come to it via one huge strongly bound connection somewhere upstream or a strong signal that had come through a whole bunch of little connections that add up to a big signal? Either way, you have this strong current flowing by, so what's the difference?

Imagine the dam again, but instead of just one big gate to let water through, imagine lots of tiny portals. When a sudden wave of water slams against the wall of the dam on one side and pushes through the portals, it comes through the other side as a fine mist. A fast moving spray instead of a stream. That's a qualitative difference in the type of flow, and it's the analogue to what I think the humour network is listening for.

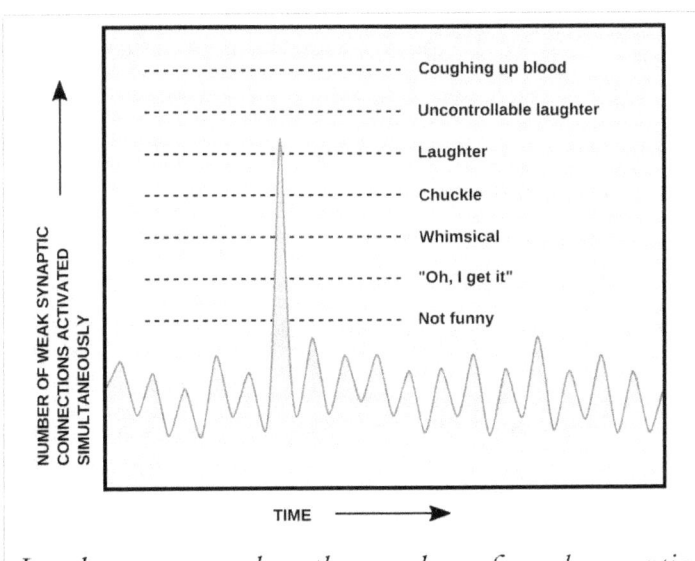

Laughter occurs when the number of weak synaptic connections activated at one time spikes above the usual noise.

Both strength and quality, mist or stream, matter for the humour network. The threshold of laughter is not a binary, on or off, thing. As we know in life, laughter can range from a slightly amused smirk up to uncontrollable hysterical laughter. In terms of the dam analogy, a few droplets is not very interesting. A thick spray rushing by indicates something big is going on. In this way, the degree of laughter response could be correlated to the amount of weak synapses activated. If we were to graph the potential for triggering the humour network, on the low end we would have a level of constant activity where the laugh centre had no response at all. After all, at any one moment there are always at least some weak synapses being activated, part of the constant random feeling around that neurons do. Above a certain minimum threshold, we would have a gradient of transition. With more *collective* weak synaptic activity, the laughter response increases. Graphed over time, there would be a constant "noise" of connections being activated as part of the normal operations of your brain, and every now and again there would be spikes that reach up into the laughter zone. The higher the spike, the bigger the response.

It might seem that once the water has rushed up to the dam, splashed through, and created a mist that activated the sensor, that we have at that moment completed a particular sequence of events. This correlates with a common concept of potential humour, particularly with deliberate jokes, that they are a distinct type of thought, with quantifiable boundaries. In our day to day experience, not just jokes, but lots of concepts seem to complete themselves, coming to identifiable conclusions that we can draw lines around. It seems perfectly logical to think of "two plus two equals..." as an incomplete notion, leaving your brain hanging, and then when we get to "... four", the concept has come to an end. It seems distinct, quantifiable, and self contained. That's a logical and useful way to look at what a math equation is. However, when we're looking at how concepts are physically running around inside your brain, the activity that contained the notion of "2 + 2 = 4" did not stop just because the way we describe it came to a conclusion. That flow just kept on trucking along, zipping through your mental network, morphing into the next thought without any distinct division between one thought and the next. There are no discreet and quantifiable units of activity that divide or quantify whole thoughts. If you had stopped before saying "4", your brain wouldn't seize up, it would just flow into other thoughts without missing a beat. We're fundamentally analogue, not digital.

Returning to the metaphor of flowing water, let's now take the dam out of the river. Instead, we have a tumultuous river with white water rapids. Constantly in motion, constantly swirling. Lots of the movement is just straightforward, part of the usual activity. As currents and eddies move around in within each other, they sometimes splash and create white water and spray above the surface. Our sensor is now a mesh of crisscrossed wires, like a net hanging just above the surface of the river. Most of the time the water doesn't spray high enough to touch it, but a big enough splash of water can. That's the most accurate analogy to how the humour network is operating. The humour network is not looking for anything to be complete, the river of thoughts underneath just keeps flowing. Every time a large white water splash happens, wherever and whenever it is, the humour network registers it and sends a signal to the laugh centre for a response.

☺ The Punchline

"A witty saying proves nothing."

~ Voltaire

We now have a physical model for what determines the success of potential humour in terms similar to Seinfeld's gap model. There is a flow of activity in your head, forming patterns that represent your interpretation of some sensory experience. That flow takes place in a structure that is yours and yours alone, differentiating how individuals respond to jokes. Potential humour that is *too close* is where the activity in your brain is flowing through well established channels, with strong connections. They flow too easily in your brain, stimulating pathways that are based on concepts you've had before. Potential humour that is *too far* takes that flow of activity down pathways that essentially lead nowhere, effectively to terminals not connected to any other networks. The concept is just too out there, and you could learn it and develop new pathways for it eventually, but for now, it's not funny because it doesn't stimulate any activity. Potential humour that is "just right" at some point in its flow hits an area of weak connectivity. Neither too familiar nor too strange.

Why would the people at the next table be laughing at stuff you don't find funny? The activity in their brain is running around in networks of neurons in patterns that is built around all sorts of context you don't have. Not merely information you don't have, but networks in their brains related to how they feel about their friends, their individual perceptions of the current environment, their immediate emotional state of mind, and all sorts of connections to stuff that mattered to them and their brains, but not to you. Where they have weak connections ready to be activated is different from you. Anything, even a simple "yeah, sure" can be funny if inspires activity within your personal framework of context for that moment. Anything can be not funny if you don't have a connection to it, nor a context to support it, within your mind.

Ultimately, the most significant implication of this model for explaining humour is that because it is based on a set of physical circumstances within your brain, it is unbound by any particular type of content. Humour based on status, surprise, culture, language, body position, smell, sight, random memories, emotions, music... you name it. If your brain can think about it, it can be funny. The type of activity, possibly of weak connections being activated in a neuroplastic model, or, hedging my bets a little, maybe something similar that can account for an evaluation of *too close* versus *too far*, can take place anywhere in the brain. As a result, there is very little limit to where the humour network can be listening, which matches observable evidence. *Anything can be funny*, and *anything can be not funny*.

If they laughed, it was funny.

Chapter Three

HUMOUR

3. Humour

☺ But... Why?

> *"Perhaps I know best why it is man alone who laughs; he alone suffers so deeply that he had to invent laughter."*
>
> ~ *Friedrich Nietzsche*

In *Star Trek II: The Wrath of Khan,* Captain Kirk tells a joke to Saavik, a Vulcan. A Vulcan is a kind of alien, if you're not into *Star Trek*. Anyway, she doesn't laugh, and remarks, "Humour, it is a difficult concept. It is not logical." It's the kind of scene you'll see a lot in science fiction, because one of the tenets of the genre is that humour is special to us as a species. We tend to feel smug and superior when aliens can't keep up with our clever witticisms.

How likely is it that humour would be unique to us and lacking in other sentient creatures in the universe? Humour is universal among humans on the planet, which seems to indicate it's a pretty important feature for our species. Contrast that with something like eye colour, where it's hard to see if any one colour would provide a selective advantage over another, and that's partly reflected in the fact that eye colours tend to mix and mingle without any one of them wiping out the others. Humour seems to have been far more key to our success, and if it was necessary for our development, then maybe it's a necessary trait for any sentient creature that wants to play on, or above, our level. It might be the case that all the aliens we meet have a sense of humour, just like they probably all have some way of holding on to tools so they can make stuff, and at least some way of communicating with each other. Knowing how to answer the question of whether or not humour was critical for our species, or critical for any species, requires answering the question, just what is this humour thing for, exactly?

There are two things that I'll take as axiomatic in approaching this question. One is that our survival as a species depended on cooperating with groups against outside threats. Individual humans are soft fleshy delicacies for predators with claws and fangs, but humans in groups are bad asses who make other species extinct. Threats also come in the form of environmental and resource pressures, and groups are better for that too. For example, a group of humans foraging for berries and sharing good finds can cover a wider area and have more chances to provide for themselves than an individual who could starve before discovering that all the good berries are somewhere else. The other fundamental assumption I am making is that it was also critical for us to develop mental systems that helped construct and confirm social standing *within* groups. A person's precise placement within a group could be life or death, depending on whose authority you stepped on, or whether or not anyone would want to have offspring with you.

In our current environment, these survival problems have been rendered moot by our technology. We can manage our environment to the point of taking it for granted, making our need to cooperate with each other in order to face outside threats something that is almost never felt outside of zombie apocalypse fiction. Also, our society has been transformed by our ability to migrate and communicate quickly over vast distances, which makes our internal group status far less clear and not a matter of life and death. These create massive distortions in how we understand and relate to our innate behaviours. Nonetheless, the systems we have once served a particular purpose, even if they aren't in accord with our current environment. It's those original purposes that need to be identified in order to see just what it is that our humour network is doing.

The particular way your neurons are linked up, and where you have patches of weak synaptic connections and where you don't, is built upon your experiences and can be said to represent who you are. No two people could ever have identically structured brains, but everyday experience tells us that it's quite common for people to share similar ideas, similar moods, similar world views. Consider that when two people laugh together, that establishes that they both had networks in their mind similar enough to each other that they both had patches of weak connections relating to the same topic. Be careful not to think that this means they both necessarily had a set of weak connections in the exact same physical location in their brain. A lot of the concepts we laugh at bring together many divergent mental processes, such as mood, interpretation, visual, aural, memory, and so on. Where the weak connections actually are might be distributed all over the physical brain in different places in both people. It's an open question as to the degree two people would have to have similar physical pathways to be able to conceive of similar ideas, but we know from everyday experience that however it happens, it happens.

Two people with sufficiently similar networks in their brain have at least some similarity in their ways of thinking, and similarity is a helpful guideline when forming social bonds. Or at least, similarity is reliable enough, and evolution is all about being just good enough to keep your species going. Animals, including humans, have many different combinations of ways of identifying potential allies using appearance and behaviour as guidelines. For humans in particular, with our big-ass brains and all the thinking we do, so much of how we behave is determined by the way we think about things before we undertake actions. That makes how people around you think critically important information for how you will be placed within any group. It would be really, really useful for group cohesion for humans to have some indication of how other people thought. If you could evaluate how close or far they were to you in terms of cognition, that would be really useful. That's basically what humour does.

Note that humour does not evaluate *relative* status, just the potential for being in the same group. It's advantageous to be connected with people above you, below you, or equal in status to you, and there's no need to expose or manipulate the status itself in order to reap benefits from commonality. In other words, a high status person telling a pun and getting some laughs gets the same benefit as a low status person. They both widen, or deepen, their social group. It is, of course, *also* useful to know one's relative status in a group, but humour doesn't have to solve *every* problem, and we possess many other ways of evaluating our varied social concerns.

Is humour the best way to go about a basic determination of similarity? Might it not work to simply have one module in the brain that lit up in response to other humans to say "friend or foe"? Might that not even work better or more reliably? Let's imagine a hypothetical system that does just that. What if you went on eye colour to know who to trust? In the world of evolution, if some humans came to trust brown eyes over blue eyes, then it's obviously advantageous for some humans within that group to have brown eyes but be untrustworthy so they could take advantage of all the naive eye colour watchers. With one mutation or genetic change that altered who got what eye colour, the whole system would fall apart. After all, clearly it would be to your benefit to be capable of saying whatever you need to gain advantage while other people mistook you for being trustworthy because of some arbitrary physical characteristic. In fact, in our world as it is, don't attractive people get away with that kind of thing all the time? I'm not just being sarcastic. Lots of research shows that we are biased toward people who are hot, and have negative feelings about people who are not. If you've ever been burned by someone with a pretty face, you have felt first hand the impact of the survival of the fittest on our social interactions.

No fixed external marker is going to be good enough, because our thinking can always stand apart from it. We need an internal evaluation method, something connected to our thoughts and feelings. Further, no *one* internal thinking marker would be good enough either. If, for example, some trust mechanism was wired up only to the parts of your brain that were dedicated to appreciation of music, then some people would find it advantageous to display an appreciation for popular types of music, but still deceive others.

With the degree of complexity of human personalities, how could any one trait establish enough commonality? In order to know if someone thinks similarly enough to you for you to put your trust in them, you need to verify their entire character. In fact, even though humour is a decentralized and holistic tool to aid the development of social networks, it's still *only one of many behaviours.* Phenotype, levels of attractiveness, clothing, facial expressions and mannerisms, involuntary actions like blushing and crying, evaluating how people treat other people... there are countless ways in which humans evaluate other humans to see how close we could or should get.

Which brings us back to aliens, and if they would or could have a sense of humour. Whether humour is a better or worse way than other methods for establishing commonality is debatable, but it's fairly clear it's not exclusive. There are plenty of group cohesion methods to choose from, and humour appears to be just one very useful tool within a toolbox of options. The more interesting conclusion is that we're constrained by the options we see in ourselves and in other creatures on Earth. It might be the case that there are other systems for thinking beings to relate to each other that we just haven't seen yet, and aliens would feel smug for possessing whatever that was, and think humans were lacking without it.

It may even be that aliens could have a superior system than humour with more reliable results. Humour's prevalence across our species strongly indicates that for a significant time in our evolution it was a powerful tool for building group cohesion, which in turn helped us thrive and get to where we are now. However, it's by no means a perfect tool, and it doesn't always succeed in that sometimes we laugh along with people we might not like otherwise. Partly this is because even under ideal circumstances trying to determine meaningful similarities in thinking is a tough problem to solve. More interestingly, though, the humour network often fails at that task because it simply isn't designed for our modern environment. We evolved the humour network over a long period of time in which the normal state of affairs was to be with same people our entire lives, walking around within a short distance of each other all the time, and never sleeping too far apart. We used to be a lot more alike and the humour network's job was a lot easier. Laughing together was probably more about affirmation of group ideas, making sure people stayed in the flow of the direction that collective thinking was already going in. Now, not so much. Our social groups are wildly varied in terms of how long we've known each other, how well, how we keep in touch, and why we are together. In that context, the humour network is trying to tell you something about similarity but falls short because there is so much about the other person that can't be assumed. Humour, as a system for confirming and creating intragroup bonds, might have been really effective when our social systems were much smaller and simpler. In our modern context, though, it has been seriously disadvantaged, so much so that it might not even have much to contribute on that front at all anymore.

☺ Don't Stand So Close To Me

> *"A wonderful thing about true laughter is that it just destroys any kind of system of dividing people."*
> ~ John Cleese

"Waaaaaiiit a second," I hear you saying. "If humour is there to help us form close bonds, then what's with all this *too close* and *too far* business? If we wanted to validate people who were part of our in-group, wouldn't the humour response be more effective if it signalled us when we were dealing with people that hit a bullseye with us in terms of thinking like we do? In other words, why would there ever be a *too close* condition?"

You ask the best questions, and this is one of my favourites, because it's a really interesting point. Let's consider for a moment the results of a system that did respond with a pleasurable sensation and vocal verification that you and the person you were interacting with were right on the same page. In physical terms, this means it's not the activation of weak connections that stimulates the humour network, it's the activation of the strongest connections, the ones where the synapses are firmly bound up in tight bundles.

It's no exaggeration to say that such a system would literally take you nowhere. Your thought patterns that you already share with the people you know would just strengthen and strengthen, and new configurations within your brain wouldn't have the same value. I wouldn't go so far as to say there would be no new learning or new ideas. Humour isn't about learning information, or verifying what's true, it's just about relating to people by sharing ideas. The ideas you share can be factually false, but at least you know you're in the loop. Also, humour aside, there would still be ways in which to learn new things. New facts and new ways of doing things would still have the benefit of potentially helping humans survive better and be passed along. It's just that if the humour network responded to the strongest connections, all the fun of laughter would be constrained to repetition of what's been done before. There's nothing that special about knowing how to cope with unchanging circumstances. Every animal that has not already been made extinct by going up in flames already knows that fire burns and is best avoided. Verifying that over and over would be pointless.

Humans do better because of our ability to cope with changing circumstances, new environments, and different contexts. A system that responded to ideas that were *too close* would only be sharing and reinforcing known ideas. In that case, in the forests of our primitive ancestry, what you found funny would be repetitious, never changing, and shared within a tight group of people that just about never varied. A situation like that might even generate too much conflict.

Research by Nancy Bell, a professor at Washington State University, demonstrated that we are harsher on people we are closer to than people we don't know so well when they tell us something that fails to make us laugh. When a stranger tries and fails to make you laugh, you merely don't laugh and don't think much of it. When someone you love tries and fails to make you laugh, you feel slighted that they don't know you as well as they should. Put into a context where humour drew hard lines between people who didn't think *exactly* alike, there would be a lot of potential for intergroup conflict, and even less ability to accept new members or assimilate with other groups.

Adaptability is good, rigidity is bad. With a humour system that operates within a sweet spot of not too close and not too far, you can *expand* your social networks by bringing in people who *might* be like you. Laughter is an *opportunity* offered to everyone in the group to include themselves. With the potential to respond to ideas that are further out from the norm, the opportunity for expansion is that much greater.

Not too much further from the norm, though. For the other extreme, of a laughter response system that was wired to respond to concepts that were *too far*, I doubt much explanation is necessary to see the problems. You can easily see that such a system would pull social cohesion apart, by making us only enjoy humour with people who had wildly different ideas about everything. As we merrily went along with people who behaved in ways totally alien to us, our group would become a rogues gallery of wildly different points of view, disagreeing over everything while laughing about it. If consensus could somehow happen in those circumstances, no one would appreciate it.

Evolution seems to have optimized the system to allow us to appreciate ideas that help us expand our thoughts while expanding our social group, but not get crazy with ideas that are bizarre to us. Not too much, and not too little. It requires constant calibration, always fuzzy around the boundaries, so it's not perfect or free from error, but it worked well enough that we all have it.

The point can't be stressed enough, though, that the wonderful benefits of group cohesion that humour might have given us in our evolutionary past are now distorted almost beyond recognition in our modern context. So much so that we shouldn't be looking too hard for examples of the evolutionary purpose in much of how humour, and especially comedy, is experienced these days. A comedian on stage is not bonding with you, not necessarily presenting ideas that are useful, not sharing a world view in order to validate anyone's idea, or anything like that. Comedy is to humour what porn is to lust, what horror movies are to fear, and what love songs are to love. Evocative and reminiscent of the feelings that we are biologically wired to have, but taken to extremes that deviate far from the goals those feelings pursue. Further, the craft of comedy feeds back into our expectations of what humour is in our daily lives, distorting our expectation of what humour can and should do for us that much more.

☺ False Positives

> *"Humor is reason gone mad."*
>
> ~ *Groucho Marx*

The kind of distortion our humour network was not made for can be seen all around us. For example, on an episode of *American Idol* a few years ago, they featured the people that didn't make it. One auditioning talent, calling himself "Jay Smoove", performed an over the top love ballad. His signature move was to hold his hand up, and at a key moment in the song, he released rose petals and sparkling silver confetti. It was completely cheesy, and the judges recognized that and laughed. I recognized that and laughed. Everyone from the show's producers to the video editors knew it was cheesy, which was why he was featured in this show of out-takes. Everyone could see that he had clearly missed the point. Except him.

That humour is part of the set of tools we use to establish a shared world view seems, on the surface, to contradict a very common phenomenon, which is when we laugh *at* someone trying to reach out to us. Not at a joke shared by two people at the expense of a third, because obviously the two laughing benefit from commonality. I don't doubt from the earliest days of humour any two people would be willing to sell out a third if it could establish a bond. The trickier phenomenon to explain is when someone is sincerely trying to relate to us, but failing at it, and being funny as a result. If it were the case that we find incompatible world views *too far*, shouldn't the socially hapless be simply reviled instead of being unintentional fodder for amusement?

To resolve this apparent contradiction, we need to remember that to our brains, the flow of activity is just that, a flow of activity, with no one flow being more or less meaningful than any other. Among our neurons, this flow is merely electrochemical signals in, and electrochemical signals out. In other words, what our brains take to be patterns of activity that can have relevance and can inspire connections is not necessarily the same thing that we would consciously identify as commonality between people. What we might describe as being the qualities that two people can share to make a meaningful interpersonal relationship comes at the end of a long string of biases built on layers of psychology, sociology, and culture, many steps above the basic operations of humour sensation in the brain.

The people who make us laugh without intending it are *trying* to be a part of a relationship. And they succeed to a certain degree because they take on all the conventions and symbols that the rest of us use successfully. Elaborate stage presentations involving confetti or rose petals have been used to achieve the intended effect by other famous performers, which was exactly why Jay Smoove tried to emulate them. His mishandling of the conventions of other performers was close enough to be in the same space as them, but far enough that we experience patterns we didn't quite expect. In other words, it was funny.

In the context of a global society, the vast array of options for social interaction is way more varied and complicated than our monkey brains were "designed" for. Given that we were built to deal with establishing a consensus on what is "cool" with a group of about a hundred naked and furry extended family members, it's no wonder that people miss the mark when trying to appeal to a nation of millions of television watchers. In the majority of our evolutionary development, not only would it be much easier to come to a consensus on what various interpersonal signals meant, it would be mandatory. In the harsh conditions of the wild, it would be imposed on your group by selective pressure to standardize your communications so as to thrive as a cohesive pack. In modern times, though, not only is it much harder to select the right signals, the selective pressure is much less. Ending up featured as a blooper in a collection of out-takes might seem incredibly embarrassing, but it's far lower stakes than being abandoned by your pack and left to die in the cold.

For hundreds of thousands of years, or maybe even millions if we include our pre-human ancestors, we constantly sought to perfect our pack cohesion through mutually agreed signals. And then, from the perspective of evolutionary time scales, *BAM!* Suddenly and without warning our social groupings are completely blown apart. I live 7,500 kilometres away from my birthplace, across terrain, mainly ocean, that would have been considered to be the boundaries of physical reality by earlier hominids. I live in a community of people who are mainly of a culture that is very different from mine, and a lot of the people I consider friends are also from equally far away places, with yet more variant cultures and world views. Every day I'm surrounded by strangers who, from a biological point of view, could be described as a completely new category of people by both not being in my pack and yet who are not enemies. The whole concept of benign strangers, like those who routinely sit around me in a coffee shop where we are not competitors for resources, is totally out of the ordinary for our species. I speak to some family and friends via the internet, some by video, and have very close personal connections to them, and yet they are in London, New York, Vancouver, and other places. It's possible now to know some people that you are almost never in physical contact with better than the person standing right beside you at a street corner. Not to mention all the peculiarities of knowing people through different media, like text, video, voice... I know people through mailing lists that I have never seen and never heard but have interacted with via text for years. You can now know people who don't know you, which is the definition of celebrity, which never existed for most of our evolution. All it takes these days is a blog and people can know something about you without you knowing who they are at all.

I could go on and on, and I already have, and still not cover all the differences between how we relate now and what we are evolved for. In that miasma of conditions around our relationships, it seems obvious to me, as I hope it is to you, that our humour response could get a lot of false positives.

It's a challenge for people to try and get across the right signals to each other, and we can all readily bring to mind instances where we missed the mark for whatever reason. For most of us, our teenage years were all about the desperate struggle to work out what those signals were, how to be in control of them, and how to read them in others. The simple act of wearing clothes comes with tons of interpersonal signal implications. Sometimes you wear a shirt and no one really thinks anything of it, sometimes you wear a shirt and everyone thinks it really suits you. Other times people think you look completely out of fashion. Sometimes, in a sweet spot amongst all the possibilities, the shirt you wear is not so close so everyone thought it was just another shirt, and not so far as to have people wondering why anyone would wear such an obviously unfashionable style. People laugh, leaving you wondering what went wrong when you were trying so hard to fit in.

As a person who not only performs comedy but runs comedy shows, I see many comedians get on stage and get laughs that are false positives all the time. It's no wonder, given all the complexities outlined above, that if false positives can happen when trying to build a sincere relationship then false positives would only be that much more likely when trying to build a relationship intended to inspire laughter. Part of the skills a comedian has to master is not only learning to get people to laugh, but to get them to laugh when intended, and not mistake the wrong kind of laughter for success.

☺ Gotta Get Me Some Of That

"If you laugh too hard, you cry. And vice versa."
~ Sid Caesar

The humour network in our brains, if it is a system for assessing commonality, would be a really useful feature for social creatures like humans. Just because it would be useful, though, doesn't mean we'll automatically get it. Personally, I think humans need wings, but I'm not flying. How, then, did the humour network get to be there, and be so ubiquitous in the human brain?

New traits can come along with random mutations, but they are almost always far less exciting than super hero comics would make us believe they are. It's hard to imagine that a robust system like humour, or even a simpler variation, could pop into existence just with some switch in a gene somewhere. It's more likely humour came about through another common way for a species to evolve a new trait, which is for there to be pre-existing features that can be exploited in new ways. The sensation of physical touch leading to pleasure and vocal response seems to be in other mammals as well as us, as mentioned in the first chapter, which would indicate that some kind of proto-tickling mechanism was present in an ancestor so far back in our evolutionary past that it is common to species as distantly related as rats and humans.

Why rats respond with an ultrasonic squeak to a particular kind of touch is beyond me. What any mammal does with their version of touch and response is their business. We also don't know why or how it manifested in our common evolutionary ancestor. However, all those questions are going deeper than this book needs to go. The point is only that evidence supports the fact that before we were even a species, there was some kind of pleasurable vocal response to touch, in some common ancestor previous to rats and people. As other species branched off from that ancestor, that touch and vocal response mechanism diverged into different manifestations, serving different purposes.

It's like how both cats and dogs have tails, but when they swish them from side to side, they have almost opposite meanings. With cats it indicates irritation or that they are about to pounce. With dogs, it generally means they're happy and excited that you just got back home after being out of the house for what seemed like *forever*. Every animal inherits traits from the species that preceded it, and then takes those features and go their own way with it. It could have turned out that humans would purr when we found something funny, or laugh when scratched behind our ears. We're asking why things are the way they are now because that's the way they turned out, not because they necessarily had to turn out that way.

We now have two essential details in place for the evolution of humour to occur. One is that the physical feature is already there in the form of laughter as a result of tickling, inherited from some mammalian ancestor long ago. The other is that there is pressure to establish commonality among ourselves within groups. Ostracism is bad, commonality is good. The question is, how did these two factors come together?

The answer is revealed in how our brain got to be structured the way it is. Think of it as a stack of blocks, where each block contains a different cognitive function. The process of evolution placed new blocks on top of old ones, so that more advanced parts are higher up than less advanced parts. At least in general. The reality is that as our brains got bigger, our skulls didn't always conveniently expand to match, so as a result, our brains are all squished up inside our skulls with not everything perfectly in order. Nonetheless, it's accurate enough to think of more primitive functions, like sensory perception and motion, as being lower down and toward the back, closer to the spine. Advanced functions, like language, are higher up and near the front of your brain, just behind your forehead.

Way, way back when we were much fuzzier, there was a tickling function, and our brains were physically smaller, not yet having developed all the layers of advanced thinking. The exact purpose of our tickling mechanism at the time is opaque to us now, buried under countless adaptations since, but in some way or another it's monitoring for the right kind of touch, and responding with a vocal response and a feeling of pleasure. All indications are that the right kind of person's touch at the right time felt good, helped you form bonds with those people, and inspired a more or less automatic vocal response to let them know about it so that they could bond back. As humans evolved and added each and every new cognitive development to their brain, there's a need to know with each and every stage of development whether that new cognitive ability separates you from the people you want to bond with, or ingratiates you with them. As our brain expands upwards, with each new layer added onto the stack, it pulls tickling's capacity for bonding along with it. In this way it starts to take something that is much more of a physical response and integrates it with more conceptual situations.

Just like it's easier to build a new house and put in all the electrical wiring where you want it to go than it is to take an old house and retrofit it with modern electrical systems, the humour network is probably more integrated into the newer areas of the brain than the older ones. Certain modules of the brain, for vision and smell and fight or flight reflexes and so on, would have been largely established previous to any tickling mechanism or laugh response. The humour network, once it started developing, could have woven its way into established areas as well, so you could laugh at a colour or smell or being startled. However, it seems likely that the humour network had more opportunity to embed itself deeply into areas of the brain that grew with it. As a result, we are potentially more nuanced and sensitive to humour that appeals to our conscious thought, language, reasoning, and so on.

Why was it that what we can now identify as a "humour network" got pulled along into higher brain layers instead of something else? Why, for example, did it have to be laughter for humour and tears for sadness? It could have gone the other way. Nothing is predetermined in evolution, and had our species zigged instead of zagged in our development, this book would be about why we cry at funny things. It's probably not entirely random, in that tears can go unnoticed and laughter pulls in people who weren't already looking at you, so there may be reasons why one existing trait got repurposed and not another. However, we'll never see the alternate universe that went the other way, so all we can do is consider how it ended up. The tickling response was there, it was a viable option, and that's the one that got taken. The important point is that the humour network grew upward along with our brains, so that the humour network as we know it now was fully established by the time we were starting to pick up pointy sticks for hunting and making our buddies laugh by using them to pick our nose.

This also implies that it was there helping us form social bonds with every step that we took toward developing more consciousness and the ability to think independently from the group. Consider that wolves might not need too many ways of evaluating commonality in thinking because they don't possess the ability to think too differently, to deceive, or to individuate themselves. I'm not saying they have no ability to do those things, just that they don't have enough of them to put their group cohesion at risk. Humans though, even at a mid point of evolution between being simpler apes and fully cognitive, are chock full of potential to deceive and go their own way. Too much ability to deceive each other and no one can be trusted and social bonds get really difficult. The humour network could have helped us stay in groups by mitigating some of the dangers of independent thought. If the humour network formed after we had evolved our fancy frontal lobes, it might have been too late.

Evolution is ridiculously slow. No matter how slow you think it is, it's slower. Which means that much like the rest of our biology that identifies us as modern humans, our humour network probably hasn't changed much since a few hundred thousand years ago, around when scientists generally agree we became the species we are now. We imagine our sense of humour to be way more sophisticated than a caveman, who would probably laugh hysterically at a well timed rock to the head. As far as I can see, though, we still laugh at a well timed rock to the head. Though it's better if we hit a guy in the balls with the rock. And if it's a baby that threw the rock, and we film it and upload it to YouTube, all the better. We still respond to really base humour, we just laugh at other stuff too, and that's probably only because we now have so much other stuff to laugh at. The mechanism of listening for the right synaptic activity doesn't need to be any different to assess the funniness of a rock to the head versus a clever and witty pun, and it almost certainly isn't, not having had the time to change too much anyway.

☺ He Who Laughs Last

"Whenever I'm on my computer, I don't type 'lol'. I type 'lqtm' - laugh quietly to myself. It's more honest."
~ Demetri Martin

If laughter is part of a system for people to develop bonds, then why ever laugh alone? If no one can hear it, it has no value. However, laughing by yourself is no huge mystery, it's simply a result of our modern environment being totally different from the one we evolved in, and us not having had the time to evolve a new way to deal with it. When laughter was formed, no human was *ever* alone. Let that sink in a bit. You were born, lived, and died, *constantly* surrounded by everyone you ever knew. In that environment, laughter was *always* a two way street, and right now your brain on the humour network level just doesn't understand that you are ever alone.

By laughing, you are in effect stating to all the people around you that you "get it", whatever "it" is that the group seems to be interested in at that moment. When you laugh, others around you have the chance to "get it" too, integrating themselves into the social environment as well. Even when you don't really get it, or only partially get it, laughing a little is better than not laughing at all, because at least then you're somewhat part of the group, which is better than not being a part of the group at all.

Laughing is not a *challenge* to people around you to check if they are on board with you. It's not a test of who is in or out of the group entirely, where those who fail are your enemies and those who succeed are your allies. It's an *opportunity,* presented to all the group's members, to exhibit their commonality with how the group thinks. It benefits both the people laughing and the one telling the joke to know who could be with them. It's not a good social strategy to draw hard lines around every person and be challenging people all the time. It's far better to try and get everyone, as much as possible, on your side. It therefor feels about as good to have people around you laugh as it does to have someone else cause you to laugh. Also, as an aside, a claim could be made that it's only important for the craft of comedy to be seen as not laughing at your own jokes, as how you act before and after telling your joke might impact how your audience gets it. Research indicates that people telling jokes among friends tend to laugh more than their listeners. In our evolutionary past, when we were among friends, not only is there no downside to laughing at your own jokes, laughter was and probably still is, the best way to get others to play along with you.

Jerry Seinfeld famously once noted that most people at a funeral would rather be in the casket than giving the eulogy. Public speaking of any kind, requires that you put yourself in a position of essentially seeking the approval of the group. It doesn't take a psychologist to recognize that for us social humans, that's a stressful situation. Standup might be the scariest form of performance because it is the most directly correlated to our internal systems for seeking group acceptance. We can easily applaud someone we think we're supposed to suck up to or sympathize with or something like that by just mechanically clapping our hands, but genuine laughter requires a level of sincerity that is much harder to fake. Not impossible, just more difficult than our more socially ritualized communications like clapping, bowing, or words. If you don't succeed in making the audience laugh, then you are not merely unfunny, you are alone. And being alone, is one of the worst fates any human can suffer.

The positive feedback from participating in the delivery of a joke could be the feeling that a lot of comedians tap into that drives them to want to make others laugh. Getting a laughter response means you are an in-member of the group. Performing comedy, like all performances, when it succeeds, comes with a high, a feeling of success and joy of self expression. It also probably comes with a sort of distilled version of this sense of having made everyone in the room validate you as part of the in-group. People who like being funny, especially those who pursue comedy as a performance craft, could possibly be the people who have a high sensitivity to this particular kind of affirmation.

☺ The More Or Less You Know

> *"The more one suffers, the more, I believe, has one a sense for the comic."*
>
> ~ *Søren Kierkegaard*

Two of my best friends, Wayde and his wife Ann, had a daughter, Senna, during the time when I happened to be writing this. I was especially keen to find out when Senna laughed for the first time, which turned out to be around five months old. At what, we have no idea, since it seemed to come out of nowhere and was directed at nothing Wayde or Ann could discern. From then on, as experienced by Wayde and Ann, and also by other new parents I know, for the first few years, children find humour in what can seem to be completely random and arbitrary things. Senna laughed when she tried to grab a slice of apple. She laughed when raised up above a certain height. The rate at which children are absorbing information, all of which is fresh in their minds and creating lots of fresh new connections, makes the potential to find just about anything funny very high. Not everything, of course. Most learning is gradual, and some things are hard to figure out. Only when experiences meet just the right conditions will they be funny, it's just that children are a little more likely to find humour in things that have long since passed into mundane for us.

This is one of two essential states for a brain to be in that can lead to finding something funny, by not having a lot of well established networks related to life experiences. This condition is by no means restricted to children. We can all have new territory to explore for humour at any age and in any context. As I learned Japanese and lived in Japan, I often found things funny that Japanese people find simplistic. For me, in Japanese, new connections were being made in broad categories. As my Japanese language ability and cultural awareness expanded, potential humour needed more nuance to inspire me to laugh. This is what happens to all of us as as we gain more experience, whether generally in our lives or in any particular field we engage in. As we become immersed in our interests and world views, we become sophisticated in terms of the jokes related to those areas of our life.

Which leads to the second condition for getting jokes, which is when there is more material available to work with because of the depth of our knowledge. As our brains develop a robust network of neurons around a particular topic, it never completely finishes and closes off. There are always yet more connections to be made around the edges and filling out gaps. It gets more and more nuanced as understanding deepens. As immersion in a topic or perception increases, the potential to find humour within it increases, but the catch is that potential humour has to be more "clever", or, more accurately, more specialized, in order to find connections that have not already been established.

The scope of this need for nuance as understanding deepens can not be overstated. It would be simplistic to think this only applies to the interests we consciously engage in. Our humour network weaves throughout our brain from the areas that cover basic intuitions and feelings up to the areas that facilitate our ability to engage in the furthest extent of our existentialist and metacontextual thinking. Not only does this mean you can find humour in moments in your life that stand outside of any particular category of interest, but that the flow of activity in your head is multidimensional to a degree that is difficult to encapsulate. You can perceive a concept from a wide perspective or narrow focus, an outside standpoint or intimately engaged, an intuitive distinction or intellectual categorization, or any other way a human brain can think. Further, there are no boundaries between our realms of cognition, meaning that our understanding of life in general could impact potential humour we get in a specific interest, and a specific interest could influence how we treat potential humour about life in general. The influences on what pathways in your mind will determine what you might find *too close*, *too far*, or funny, are infinitely recursive.

This is a large part of the reason why attempts to explain humour based on the content of jokes will ultimately fail. If you say humour is based on status, or "benign violation", or any other understanding of how a joke is constructed or presented, then a human brain can understand that metacontext for the joke and become used to it in a way that fails to inspire the humour network. The details of the content matter less if the overarching idea is manifest in the brain by well established pathways. It's like playing a card game over and over. I may not know what hand I'll get dealt next, but I can still get tired of playing cards.

On the other hand, though, the multidimensional capacity of our thinking makes absolute declarations impossible. Some people might *love* playing cards and just about never get tired of it, because their emotional engagement with it helps them to create synaptic activity in more nuanced, subtle corners of the overall pattern. Nuanced and subtle corners that I'm not interested in exploring. They don't have to know more about cards, they only need to feel differently about it. Who we are is just as much a part of the context as is the context of the thing we are thinking about. Which means, presented with the same potential humour, we can have wildly different appreciations for it. This can lead to considerable divergence, even fights, between people about what is funny.

☺ I Don't Get It

> *"If you wish to glimpse inside a human soul and get to know a man, don't bother analyzing his ways of being silent, of talking, of weeping, of seeing how much he is moved by noble ideas; you will get better results if you just watch him laugh. If he laughs well, he's a good man."*
>
> ~ *Fyodor Dostoyevsky*

Matt Stone and Trey Parker, the creators of the long running animated series, *South Park*, once satirized another show, *The Family Guy*, by saying that every joke on *The Family Guy* was a simple formula of combining of a pop culture reference, a noun, and a verb, selected at random. By manatees. Never mind that last part, though. The point is that the criticism of *The Family Guy* was that their jokes all followed a predictable format. Which I generally tend to agree with.

However, the unavoidable implication of stating that a show has a predictable format for creating jokes is that it seems to imply one believes that the people who watch it and laugh are lacking the ability to perceive that formula.

Are people who watch *The Family Guy* any dumber than anyone else? The show was saved from cancellation in its early years because of fan support, and has gone on to have twelve successful seasons and won Emmy awards for comedy. It's not a show lacking in popularity, so does that make me, and the creators of *South Park*, like the one guy at a comedy club who thinks it's the other eighty people in the room who don't know comedy? Or that the fans are somehow dumber for not seeing the obvious repetitions?

Only if I assert that the ability to appreciate metacontext the way I personally do is a measure of intelligence. Which it isn't. Knowing that humour is just a measure of the degree to which some mental activity activated enough weak synaptic connections in someone's mind at the right moment, and nothing more, I can divorce laughter from any need to evaluate intelligence. Or morality, or ethics, or personality, or anything else.

I've watched episodes of *The Family Guy* in Japan with my Japanese friends, and for them it's a very different experience from me. I may not know what pop culture reference is coming up next, but I am over pop culture references as a source of comedy in this context. Most of it is *too close*, though now and again it will jump enough outside my expectations to be just right. Not often enough for me to want to watch the show regularly, but it does happen. My Japanese friends, though, don't have the exposure to western comedy the way I do, so for them it's a completely surreal experience. Maybe they can even see the rhythms of the show as I can, but for them it's a matter of strange characters with inexplicable catch phrases popping in and out of frame to do strange activities, without much supporting context, so it's almost abstract in presentation. It certainly doesn't make them any less intelligent as people to see the show that way. A lot of the material is *too far*, but some of it hits home just right, enough that they're fans of the show. As a fan of old *Monty Python* sketches, I certainly can't critique anyone for appreciating surreal comedy. By extension, anyone watching the show, regardless of background, can come to appreciate any aspect of the show that works for them. Just like I can walk away from a card game that I've become tired of, that doesn't make the people who stay at the table for another round any less perceptive.

If you like *The Family Guy*, or anything else, you're no dumber than the person who finds it obvious, and no smarter than the person who doesn't get it. You merely have different pathways in your mind built around the material being presented. Humour is not a measure of intelligence, full stop. If you like *The Family Guy*, then all we can say about you is that you're somewhat in the same zone as the people who create it. And even then, only in the same zone for being at play with certain ideas, in a particular way, within the context of that show. No more, no less. We can laugh at things together in spite of our brain's potential for divergence from each other because of our brains default disposition of wanting to try to come together. With our modern society being as complicated as it is, though, our ability to find commonalities in more specific and contained ways becomes ever more refined.

With all the modern social conditions that limit the humour network's value as a tool for assessing commonality, people can easily find themselves laughing at things they would otherwise consider immoral, or stupid, or laughing along with people they wouldn't otherwise get along with. When someone beside you in a comedy club laughs at the same jokes as you, is that really telling you that much about that person? Not only is humour removed from its original purpose as a result of the natural progress of human civilization, people engaged in the craft of comedy in all mediums are frequently *deliberately subverting that purpose* by challenging themselves to get audiences to laugh at things the audience would not usually consider funny. Comedians want their material to stand apart from the run of the mill jokes shared between friends on a day to day basis. Comedians want to be stage-funny, not merely friend-funny, which results in the craft of comedy extracting itself further away from humour's purpose, and taking our expectations of what it means to be funny along with it.

Comedy is a whole different thing from humour.

Chapter Four

COMEDY

4. Comedy

☺ Two Types Of People Walk Into A Place And Say A Thing

> *"A pun, a play on words, and a limerick walk into a bar. No joke."*
>
> ~ *Unknown*

If anything can be funny and *anything can be not funny*, the only way we know what inspires the humour network is the evidence of laughter. However, everyone has had the experience of hearing something that they are sure was intended to be a joke, but it didn't make them laugh. What is it we're identifying when we've got something that is not funny, but we know it's *supposed* to be a joke?

Although *humour* is only confirmed with the presence of laughter, *comedy* can be shaped by cultural expectations, media, personal habits, and so on. In English speaking cultures, there are whole catalogues of jokes that begin with standard set phrases like, "A guy walks into a bar..." or, "A guy goes to his doctor..." and that sort of thing. When we can see how something was supposed to be funny despite its failure to make us laugh, we are recognizing a pattern of comedy *style* familiar to us. The "guy walks into a bar..." preamble to jokes is an obvious example, but our sense of what is intended to be comedy is much, much more nuanced than that. There does not have to be any specific conventions at play in order for someone to intuit that an attempt at comedy was made. All it takes is to perceive the way in which potential humour feels similar to contexts that usually inspire laughter.

The craft of comedy isn't just an entertainment option built upon our capacity for humour. It feeds back into our daily lives and shapes our expectations of what humour is. Which is not unusual among art forms. For example, maybe more now than ever before in our history, pornography has shaped a lot of what people think sex should be like. When a comedian gets on stage and tells a joke that makes a room of hundreds of people laugh, it's hard not to feel that there's something about those jokes that are funnier than the jokes we tell amongst our friends. Even though your individual output of laughter could be equal in both situations, the validation of so many more people responding to the comedian is very compelling. We take cues from how comedy is presented, whether from live performances or via media channels, and introduce them into our daily lives in hopes of having more satisfying experiences.

That there is any particular "right" way of being funny, or a better way, or a more successful way is an artificial construct, and ideally shouldn't be confused with the capacity for humour that everyone can participate in.

☺ Back Up A Sec

"In risu veritas." (In laughter, truth.)

~ James Joyce

While you often hear people say "it's funny 'cause it's true," you pretty much never hear people saying, "I didn't think that was funny because it was made up." Lots of jokes are based around entirely fictitious, surreal, and exaggerated descriptions. Similarly, while you often hear people say, "I didn't laugh because I was offended," you don't hear people say, "I laughed because it was righteously moral." Quite the opposite, as many people consider jokes that are too safe, or "politically correct", to be too tame and lacking the necessary edge to be funny. Both instances share something in common, which is that they presume that the foundation of the success of potential humour was built upon an evaluation of the content, which we already know is problematic, and hopefully I've beaten that horse to death sufficiently enough to not have to retread the issue.

Still, both truthfulness and offensiveness inspire some really strong gut feelings in people. When someone feels that a joke has revealed a truth, they can be awestruck at how comedy is a tool for insight. When someone doesn't laugh because they think a joke is too offensive, they might find it literally inconceivable that anyone could think the premise was funny. We don't always associate comedy with morality or truth, as we acknowledge there's nothing particularly true or ethically challenging about a pun or a surreal prop gag. When truth or offence comes up, though, the feelings behind the association can be really strong, firmly convincing us that these are qualities interwoven with our reasons for having laughed or not laughed. What is the source of this firm conviction?

One of the fundamental ways in which your brain thinks about things is to look for reasons and explanations for everything that happens. So much so, that even when you didn't actually anticipate a cause and effect relationship between two things, when you think back on it, you will tend to naturally assume you knew the cause in advance of the effect. This is known in cognitive studies as a "hindsight bias". Or *vaticinium ex eventu*, because saying it in Latin makes it seem more official somehow.

In our evolutionary past, if a human saw movement in nearby bushes, it made a lot of sense to run away because one should always assume that there was a tiger prowling just out of sight and about to pounce. It was much more likely that it was just the wind blowing through the leaves, but that's okay, because all the times there was no tiger were more than compensated for by the fact that once in a blue moon you avoided a fatal tiger ambush. Even if it was only one tiger out of every ten thousand times there was a rustling in the leaves, it was *always* the best strategy. Jumping when it was just the wind only cost you the tiniest bit of energy. Not jumping that one time when there was a tiger meant death. It's like a lottery in reverse, where the odds of losing are very low, but the stakes are fatally high.

Every time a human jumped, they tell their friends, "I heard a tiger over there behind the leaves, so I jumped." Well, they might screech and wave their arms around in a panic because this happened as much before language as after. Still, they wouldn't have jumped if they thought it was just the wind, so, in order for a primitive creature to maintain the better strategy of always jumping, they always *believe* there definitely was a tiger. The times when there was no tiger is not evidence of anything contrary, because, the primitive human reasons, you didn't see a tiger as a result of having jumped away. The tiger, missing her opportunity to jump out and catch you by surprise, walked away without ever revealing herself. Thus, in all cases, it makes evolutionary sense for the human mind to accept as fact, "I jumped because I heard a tiger. Definitely a tiger. No two ways about it." In contrast, a robot, constrained by logic, would have to admit, "I thought I heard what could have been the result of the presence of a tiger, but it is unclear what actually happened." The whole reason we are here today is because that jittery and paranoid way of thinking that was usually wrong about the actual facts works really well in the life or death struggle of nature. It's known as an "adaptive bias", favouring evolutionary success over objective accuracy. We are the descendants of people that jumped at every rustling noise in the bushes, and the robots that just sat there, all smug in their logic, got mauled by tigers.

It's still largely an okay strategy, because when you hear a crashing noise behind you and you turn around to see some broken glass on the floor, the odds are good that the sound and the aftermath are connected to the same instance of a cup falling to the floor. Usually, there aren't any downsides to that kind of assumption, but it can be problematic. There are countless insurance claims and court cases everywhere where people drive cars, where it is disputed who hit who first because the "witnesses" were people who heard a crash, *then* turned around, and "knew" which cars hit which first. The problem is not just limited to simple mistakes of sound and sight, either. In our modern environment, we encounter more and more circumstances where so much of what we expect of our scientific technologies and social politics relies on more objective analysis of statistical information. Statistical information and probabilities that can lead to counter intuitive conclusions. As our understanding of the universe, our world, and ourselves, becomes continually more sophisticated, so does our need to be just as sophisticated in how we understand it. Unfortunately for us, though, our brains can't evolve fast enough to keep up with the pace of our social and technological development.

How all this relates to potential humour, particularly offensive material and jokes that seem "true", is that we need to take it as axiomatic that our brains are wired in a particular way for reviewing the past that can be misleading. We routinely assign causes to results as if we knew them all along, when the reality is the order of how the information came to us doesn't match our memory. Not only is this backwards rationalizing misleading, it's so deeply ingrained in us that we feel strongly in our gut that these miss-ordered conclusions are correct. It can be hard to convince us otherwise.

For comedy, this means we can, and *almost always do*, draw the conclusion that a joke was funny *because* of the way in which we understand that joke *after* all the activity that represents that joke has already been evaluated by the humour network. However, now that we can distinguish the humour network's response to neuronal activity from the content of the flow of activity in your mind, we can see that laughter and disposition in response to an idea can be two different things. Your thoughts are always flowing, never stopping, so that after you've experienced potential humour, whether you laughed or not, the information representing that potential humour is still there, still flowing around the pathways in your head, and you'll think about it no differently than how you think about anything else. You'll evaluate it for whether or not you think it's true, or offensive, or boring, or cute, or insulting, or informative, or whatever. In the case of things that strike us as true or offensive, they can resonate with our sense of reality and justice so strongly because those are areas to which we frequently have strong sentiments attached. So strong, that we let them colour our interpretation of how we came to have those feelings, and the meaning behind our reactions.

Consider that so many things are true and not funny, and so many things are funny but not in any way true. There is overwhelming, and easily observable, evidence of the fact that truth has no reliable connection to why we would laugh. Still, when something comes along that is both funny and true, at least as far as we're concerned, we have absolutely no problem asserting, "it's funny *because* it's true." It just feels so right, so comfortable, because making that kind of leap of logic is pretty much exactly what our brains are designed to do.

We could say, at least in this context, that there is a three step process when receiving something potentially humorous. First, we take it in, and the experience inspires a flow of activity in our brains, zipping around our existing pathways. Second, at some point, that flow might be right for triggering the humour network, and maybe we laugh. And third, now that we have the entire joke, all the information, we evaluate it and form opinions just like any other information. The tricky part is we're built to perceive this order of events differently. We tend to be sure we did step three, evaluation, before we did step two, laughter. Especially when it comes to information that hits home on sensitive topics.

☺ You Can't Make Me Laugh

> *"It is the ability to take a joke, not make one, that proves you have a sense of humor."*
>
> ~ *Max Eastman*

Sarah Silverman, in an interview with Conan O'Brien in 2001, told a story about how a friend of Sarah's advised her that one can avoid jury duty if you make it known you're obviously racist. Later, when Sarah was being considered for jury duty, she had to fill out some forms. She thought she would follow her friend's advice and get out of it by writing something obviously inappropriate, like, "I hate chinks." But Sarah didn't want to be thought of as *actually* racist, so instead she wrote, "I love chinks." Sarah's anecdote generated some controversy. A representative for the Media Action Network for Asian

Americans, Guy Aoki, said her use of the term "chink" was not successful satire.

Can Mr Aoki, or anyone, make objective judgements about whether or not something is "successful" satire? With humour being just a measure of the level of a certain activity in the brain, it is not a reliable indicator of anyone's ethics or morality. At the same time, it could be argued that even if everyone thought Sarah Silverman's joke was funny, it might still not be the kind of joke we want to pass around in society.

The Sarah Silverman versus Guy Aoki story reminds me of something that happened with some friends of mine, a long time ago, when the comedian Andrew Clay Silverstein, better known by his stage name Andrew Dice Clay, was having his fifteen minutes of fame. My friends got together to watch Clay's standup special *No Apologies*, which came out in 1993. My friends were in their late teens or early twenties at the time, and they thought it was hysterical. Right up until somewhere around where Andrew Dice Clay starting talking about how, "Chinese newscasters don't act Chinese." All these friends happened to be Canadians of Chinese heritage, and they didn't laugh.

They had sat through racial slurs, misogynist statements, and general bigotry without thinking it was anything other than harmless and intentional pushing of boundaries for comedic effect. Were they less sensitive to the plights of other minority groups? If you were to discuss the matter with them, they would put all words of racist implications on the same level playing field of offensiveness. They don't think "chink" is any worse than "spic", or "kike", or "nigger", or whatever else. However, they have more depth of association with the word "chink", as they had more neuronal networks built upon their experience with it. More experience means more pathways in their minds, which means that particular word, and racist sentiments against their particular heritage, are *too close*. Closer than the other concepts, even though they objectively understand all bigotry to be equally bad. When Clay started touching on a topic that was *too close* in their minds, they evaluated the reasons why their opinion of Clay had changed mid act, and assumed that it was because *now* Clay had crossed a line. Which is what we can expect, as a result of the hindsight bias that all humans share. If you look at Clay's performance objectively, though, he was walking the same line all along.

When a comedian is telling you a joke, if you are enjoying them, you are largely *allowing* them to lead you along. You can imagine in a way that you build the rails, and the comedian drives the train. However, if something happens to make you switch rails, from that point on it might be very hard for the comedian to get back on the track that they want to be on. The flow of thoughts now includes some new pathways which impact the potential for the right synaptic activity. For my friends watching the video, the new opinion they had of Andrew Dice Clay had completely shifted the activity in their minds so that Clay was no longer an edgy comedian at play with terms that offend. Now he was just a bigoted jerk, spewing out one meaningless attempt to shock after another.

Similarly, Guy Aoki didn't find Sarah Silverman's joke funny because, even more than my Chinese Canadian friends, as a person who makes it his life ambition to fight discrimination against Asians in America, the word "chink" lands in a bed of neural pathways so deeply embedded in his mind that it would be very, very tough to trigger the humour network. Once he went through the experience of hearing Sarah Silverman's words and not laughing, the hindsight bias led him to believe he had not laughed because she had done something wrong. Whether or not it is wrong for someone to throw around the word "chink", or any slur in comedy, whether ironically or not, is a debate for society to have, and I can't address that here. It's Mr Aoki's right to question whether or not the word "chink" should be used at all, for any end goal. All I can say is that Guy Aoki's assessment that it wasn't "successful" satire doesn't hold water. People laughed, indicating that for them, it succeeded in the goal Sarah Silverman set out for her joke. That Guy Aoki didn't laugh doesn't contribute anything to the debate. Nor is the laughter of anyone who did laugh. In any situation of evaluating the ethics of a statement, that it made anyone laugh or not is irrelevant.

☺ Fuck

> *"You know the seven, don't ya? That you can't say on TV? Shit, piss, fuck, cunt, cocksucker, motherfucker, and tits."*
>
> ~ George Carlin

Every comedian knows that peppering an act with some swear words can heighten the tension of the audience, and get a few laughs. There's a noticeable difference between saying, "it's hot outside," and, "it's fucking hot outside." It's like a spice that can add just a little extra kick. Every comedian also knows, or should know, that it has diminishing returns. Like making a meal out of spices and nothing else, an act that relies on vulgar phrasing without any substance underneath will ultimately leave an audience underwhelmed.

Comedians refer to this as "working blue", a term which isn't exclusive to using swear words, but also to touching on vulgar topics in general. Insofar as it refers to using swearing for emphasis on topics that otherwise might not be particularly vulgar, some comedians think it's a sign of weakness, an inability to hold the audience's attention through the strength of ideas alone. Other comedians disagree that it reflects on the material, but whether comedians are for, against, or indifferent to working blue, no one denies that swearing adds a little kick to an act. The only debate is whether or not that extra kick is a respectable choice. Something I can't answer, as that's to do with social ethics, not comedy.

The idea that humour is a matter of measuring synaptic activity in the mind also supports the observable evidence that swearing adds a little kick. In a sense, any choice of some words instead of others will inspire the flow of activity in the mind of the audience to go in slightly different directions. The addition of the word "fuck" into a sentence will involve just a few more pathways than the sentence would have had otherwise, and those extra pathways might be connected to the overall flow by some tenuously connected synapses, contributing to the activity that the humour network responds to. The problem, though, is why should certain words matter more than any others? Why should "fuck" or "shit", or whatever the taboo words in the comedian's language are, have any special significance over any other words? Even if you aren't the type of person to throw those words around casually, surely it's no surprise that other people do. They may be "bad" words, but are they really *uncommon* words, unfamiliar enough to bump a sentence up from *too close* to funny?

Taboo words may or may not be common, depending on who you hang out with, but what they aren't is equal in your mind. Research has shown that different words can trigger responses in different parts of your brain, and taboo words connect to parts of your brain buried just a little deeper, in your more primal limbic systems, where a lot of your emotional state of mind is determined.

People in brain scanners have been shown to respond to taboo words the same way they respond to angry and emotional faces with similar parts of their brain being lit up. It's not the meaning of the words, but that as a society and culture we have essentially designated certain words to be a verbal equivalent of more primal and physical displays of emotion. Thus, "cunt" is more jarring than "vagina", "shit" affects you more than "feces", and "fuck" has all sorts of impact on people in a way that "intercourse" never can.

The different pathways in your mind that get inspired by taboo words aren't adding a kick by merely involving arbitrarily different pathways in your mind the way that all word choices do. The kick they add comes from adding different pathways in an emotional part of your brain. It's more like the difference between whether a comedian screams or whispers the words they say, or says them with an angry or happy face. Certain words simulate those emotional contexts.

Of course, that emotional kick that taboo words can provide by no means guarantees a better response from the humour network, just a higher emotional response, which may or may not help. If Chris Rock gets on stage and talks about "niggers", he is more likely to make it work for him because the emotional context is accepted within all the rest of the context that he brings to the stage with him. A white middle class performer like myself would have to work much, much harder to create a context in which I wouldn't simply be left dealing with a very non funny and emotional aftermath. In other words, the kick that comes from using taboo words will bump up ideas that are already, or close to, being funny on their own. If the ideas are too lacking, then taboo words will still raise the emotional stakes, but very possibly against the comedian.

☺ Comedic Uranium

> *"It is well known that humour, more than anything else in the human make up, can afford an aloofness to rise above any situation, even if only for a few seconds."*
>
> ~ *Viktor E. Frankl*

For years I have run open mike nights, where new comedians come and try to see what it takes to be on stage. As a result, I have heard more attempts to make jokes about rape and paedophilia than any human should have to endure. The vast majority of attempted "shock" humour fails to make people laugh, and I've seen it provoke outright hostility from an audience. Are there some topics that are just too beyond our cultural or psychological norms that they can never be funny? Cancer, rape, genocide, paedophilia... Some people, especially those with personal experience with them, might find it hard to believe they could ever be made funny. At the same time, there are successful comedians who have made careers out of finding the humour in dark topics. The whole reason so many beginning comedians try so hard to be "shocking" is because they are emulating the comedians they admire who succeed at it.

It's been described from a few angles now how humour is not a measure of morality, or anything else, just what's funny, so it's no surprise that comedy could be made out of seemingly unapproachable topics. A question remains, though, as to why those darker topics would have any more appeal to comedians or audiences? If they had equal potential to cause laughter as any other topic, then it would make much more sense for comedians to avoid them, as they could get the same amount of laughter from less controversial topics, and not suffer any risk.

Just like taboo words, controversial topics involve a certain emotional energy, but they aren't as blunt as the utterance of a single "fuck" on stage. Sometimes it takes many sentences for a comedian to express what it is they are talking about, meaning the activity is more multifaceted than a simple knee jerk reaction from the limbic system. The presence of a broader taboo topic has a lot more going on than just adding a little kick to a comedy routine.

Could it just be that there are additional benefits, aside from laughter, that a comedian reaps from making a taboo topic work? Perhaps it's just the case that when comparing comedians, even if two comedians got the same level of laughter from an audience, the one who did it by using taboo topics seems more "clever".

I often describe topics that risk being too extreme for public performance as being like uranium. Mishandle it and it explodes and everyone gets sick and dies. But, if you treat it just right, it can power a city. I don't think that in itself is a particularly brilliant insight, though. I think a lot of people who know comedy know that the art of making taboo subjects funny is the way in which one rides the razors edge between funny and unacceptable. However, by putting it into the context of brain activity, more light can be shed upon what exactly is happening that separates funny from disastrous.

Uranium topics are, like every other topic, just another flow of activity within the pathways of our mind. Every pathway in your brain is ultimately connected to every other pathway, but not everything is equally accessible. Some pathways will have more limited routes, with less neurons and synaptic connections involved, depending on how you relate to the information that goes through those areas of your brain. "Uranium topics" not only exist in areas of limited physical connectivity in our mind, but in addition, the only routes to them are associations with negative feelings and imagery.

Compare for a moment the charged word "rape" with the mundane word "work". You can simultaneously love and hate your work, have all sorts of memories about it, the people you met there, your aspirations for what you want to do in the future. There is no way I could list all the ways that a concept like "work" can have different associations, which in effect means it can branch off into many different patterns of your mind. "Rape", however, unless you're a sociopath, has a much more limited scope. It's never a good thing, it's hopefully not something you've ever had direct experience with, nobody wants it, hopefully it's not in your future, and so on. It's still a multidimensional and complicated topic, but you can see that it is arguably much more limited in scope than a topic like "work".

The actual number of patterns of neuronal connections in your brain that apply to the concept of rape is much lower than other concepts. Because the number of associations is small, that means the potential for different people to have associations in the same way is more limited. Making a joke about work has a higher chance of success because there is so much that you and I might share on the topic. So many more potential associations. On a limited topic, the chance that we have shared patterns, close enough that we can mutually empathize, is simply less. The more difficult the topic is, the less likely it gets.

Not only are the connections fewer, the connections that do exist are so much more likely to be bound to something negative. If you tell a joke on a taboo topic that doesn't work, then in the mind of the audience you are probably left hanging in the pathways where the negative associations are. The audience thinks about what you just said from the perspective of being in that negative space, and they assume that your attempt to be "shocking" was the cause of your failure, as per the hindsight bias. Really, though, it's not that you brought up a topic that could never be funny, it's that you didn't properly account for how the audience relates to it. You just didn't navigate the pathways successfully. Which is really no different than the risk involved in telling any joke, just that your perception of the audience's position has to be better.

The corollary to all this risk is that if you can successfully navigate through the negative associations and bring the activity in the mind of the audience to a remote network of neurons, you can be rewarded with more laughter than you might have received with other topics. Precisely because of the limited scope of the neuronal networks associated with taboo topics, they have a high likelihood of having lots of weak synaptic connections, ripe for activating. This is the gold that comedians dwelling in "shock" humour are mining for. It is true that if you can get there, the laughs can be very robust. Further, the hindsight bias can now work in favour of the comedian, as the audience will now assume that they laughed *because* the comedian was more clever for having got them there.

☺ Would Comedy Exist In Utopia?

> "If you're not allowed to laugh in heaven, I don't want to go there."
>
> ~ *Martin Luther*

In the 2006 movie *Borat*, Sacha Baron Cohen, playing the title role, takes a comedy class where he asks the instructor, in his character's accent and grammar, "do you ever laugh on people with retardation?" The instructor, a public speaking coach named Pat Haggarty, replies that, "here in America we try not to make fun of, or be funny with, things that people don't choose." Cohen responds, "But perhaps you have not seen someone with *very funny* retardation," and then goes on to tell a story involving his sister being raped by his "retarded" brother. Which gets a laugh out of Haggarty, but nonetheless Haggarty advises the story wouldn't be funny in America.

To be fair to Haggarty, he's probably laughing ironically at the metacontext of how inappropriate the story is. But then, isn't ironic metacontext a perfectly valid way to make something funny? As viewers, we know Sacha is making up the story, and deliberately provoking the teacher, so if we laugh, and I did, it's not technically making fun of "retarded" people, but it is using the topic for humour. What does any of this prove? I don't know, I get kind of lost trying to work out all the layers. I do know, though, that a lot of people would definitely agree with what this humour coach was trying to get across, which is that it's okay to make jokes about things that people have control over. It's alright to poke fun at someone's job, or the choices they make in who they date. It's not okay to make fun of someone for their race, or a handicap, or gender. Fair enough. Although, again, I'm not sure it's easy to work out the boundaries. The line between choice and a circumstance is not always so distinct. A job might be something that someone was compelled to do because the socioeconomics of where they come from left them little choice. Dead hooker jokes, anyone?

When we are confronted with someone who comes along and tells us we shouldn't use the word "retard" in our jokes, because they are "people with special needs", we might feel like our fun is being spoiled. We know we didn't actually mean anything hurtful, we're just being "edgy" or "ironic". People get upset that political correctness kills comedy, because it subtracts from what can be made fun of. Once we've made everything equal in human society, once we're out of disparities and struggle, then there's no more material to work with.

That's a false dilemma. If you're cynical, you might comfort yourself with the idea that human society will always have its disparities, and so there will always be a place to draw comedy from. To look at it that way, however, is to limit yourself already, because you've narrowed yourself to drawing comedy from the kinds of relationships that you can conceive of right now, and discounting possibilities you simply haven't experienced yet.

What if society were to become completely equal? What if it was heaven on earth, and we were all walking around in togas and frolicking in gardens? There would still be comedy. Humour is about relationships, which will always exist, and will always have infinite variety. Comedy is merely a performance at play with the sensation. There is no reason we can't have comedy that is cynical, ironic, even distasteful or vulgar, but actually believe in, and act upon, values of social justice. Laughter is not a gauge of what is right or wrong, or even what the comedian or audience believes is right or wrong, just that we can make the same mental patterns together.

Humour will certainly exist wherever there are humans, no matter how utopian or dystopian their reality is. Whether it's cynical, safe, or even just completely different than anything we have now, humour will be a way for us to bond with laughter. In fact, since everyone gets along in utopia, and humour helps us bond, you could argue that there might be *more* comedy there.

People who bitch that their fun is being ruined by those who advocate "political correctness" are painting their straw men as having less of a sense of humour, which makes them outsiders and therefor bad. It's a counter-judgement within a debate that reveals nothing from either side, except the obvious point that different people find different things funny. Should we still be allowed to laugh at jokes about things that we wouldn't say in earnest at any other time? I don't know, that's an ethical issue beyond the scope of humour. All I can say is that political correctness is not the antithesis of humour, but nor should it be the judge.

☺ Giggling At Funerals

> *"Life does not cease to be funny when people die any more than it ceases to be serious when people laugh."*
> ~ *George Bernard Shaw*

There was a case in London in 1931 of a guy named Willy Anderson who started laughing uncontrollably at a funeral, and died two days later. It's believed that he suffered a severe "subarachnoid hemorrhage", which is a kind of stroke caused by bleeding in the brain. Maybe it rattled a node in his humour network, like the girl laughing at pictures of horses while being operated on, or maybe it was more directly putting pressure in the area of the motor cortex that drives the laugh response. We'll never know for sure.

That dude wasn't the first to laugh at a funeral, though, nor the last. Most everyone has an anecdote of feeling a compulsion to laugh at a completely inappropriate time. Fortunately, it's almost certainly not the case that people are having mild strokes. Laughter comes up at funerals, serious business meetings, religious rituals, and any other place where the participants are expected to be serious and sombre because the humour network cares much less than you do about what it means to be there.

Your sense of humour is not bound by social constraints. Any social ritual humans engage in is a construct of the process of civilization that came along long after the humour network had evolved into us. As such, our humour network won't avoid finding things funny because of what we might consider "proper" behaviour. The humour network is only concerned with the activity present within neurons. If something happens that triggers a rapid cascade of weak synaptic activity, you will feel like laughing. Doesn't really matter what sets you off. Maybe someone farted and only you heard it. Maybe you farted. Maybe you just happened to remember something kind of amusing. Whatever it is, if it's sufficient to trigger the humour network, then that's what happens.

So, now, there you are at a very serious funeral, and something, some idea you just had or something you just noticed, struck you as funny. What happens at that moment is that the context of being in a place where you're not supposed to laugh feeds into the flow of activity in your mind. Then, if you try and suppress the laugh, the fact that you are suppressing your laugh also becomes part of the context. A cycle starts developing, and the humour network doesn't need to know why laughing at a funeral is bad, it just needs to know that with every pass of the flow of activity, there are layers of context being added, bringing in more waves of synaptic activity.

Whatever the context, it is precisely because of these cultural agreements that certain topics, like death, are serious that make it a ripe ground for comedy. Similar to how taboo topics can be funny. It's our deliberate attempt to make them holy ground that limits our pathways related to the concept, and keeps the potential for weak synaptic connections high. When someone finds something funny within the context of a death or funeral, it's not that they are laughing at the death or funeral itself. The person laughing at the funeral isn't less sad than anyone else, it's just that their concept of appropriate behaviour is a new angle from which to find funniness in things that would ordinarily be mundane.

If you're looking for help on how to avoid potentially awkward and inappropriate giggles at funerals, you might make the feeling go away faster by allowing yourself to think through the idea that struck you as funny. The more you think about something, the more the pattern gets assimilated and becomes mundane. By trying to not think about it, you prolong the time in which the thought fills out weak synaptic connections with activity. You can't un-think something once you've thought of it, so your only option is to try and get used to it as soon as possible.

☺ Roots

> "Everybody laughs the same in every language because laughter is a universal connection."
> ~ Yakov "Smirnoff" Pokhis

For a moment, think of all the pathways in your brain as coming together to form a shape like a tree. In this case we're just thinking about the tree as you see it above ground, starting at the trunk and working your way up. The trunk is made up of pathways that are fundamental to your identity, that aren't likely to ever change. Things like "my mom raised me" and "don't put your hand in fire". The branches are more current, but still form rather fundamental patterns. Concepts like "I am in my mid forties" or "I am an aspiring comedian". They can change, but they aren't likely to change in an instant. The leaves of this tree are your current thoughts. They come and go easily. Things like "I want a coffee now" or "That person is looking at me".

Culture makes a lot of its impressions in the trunk and thick branches of the tree. You grow up in a cultural context, and it influences who you are as a person right from the get go. Culture doesn't dictate who you are, as people can find themselves dissatisfied with the cultural norms they grew up with. Nonetheless, everybody grows up having to orient their identity around the acceptance or rejection of ideas within their culture, meaning the strands of neuronal patterns that represent culture are deeply entwined in our minds.

From that basis, we can see culture's influence on comedy. It creates a basic context in which potential humour is perceived. But the difference in seeing culture as just another set of pathways, as opposed to making culture a foundation for comedy, is in the mutability of the mind. Ultimately, you are capable of understanding concepts that any other human is able to grasp. It may take some time, training, or determination, but anything another human can conceive of can also be conceived of by you.

How much you take on new ideas so that you can appreciate humour based on those ideas is only a matter of how willing you are and how much time you expose yourself to the right contexts. After all, it does not matter whether the pathways are created recently or not, it only matters if there are sufficient weak synaptic connections currently available for activation. Culture is not a box to be constrained in, but a platform to start from.

A comparison could be made with cuisine, which is something that is founded on a biological reality common to all humans, namely that we all need to eat, but what people prefer to eat is shaped in a large way by the culture they grew up in. People give names to cuisines that bind them to their culture, so there is Japanese cuisine, Italian cuisine, Mexican cuisine, and so on. Comedy is a lot less easy to define than particular food dishes, but people do the same by saying there is Japanese comedy, German comedy, British comedy, and so on.

Sometimes, people who grew up on one cuisine will not recognize other people's cuisines as food. A lot of Americans look askew at eating chicken feet in a Chinese restaurant. I went to Korea and was not about to try eating the fried silkworms one guy was selling at a road side stall, although my other friends, one Canadian and one from the US, were all about trying new foods. That I was not willing to try a new cultural food but my friends were, even though we were from the same culture, is a good analogy for culture and comedy. Our culture was a factor in how we collectively approached the question of whether or not we considered this cuisine to be "food", since we all considered it weird to eat bugs. As individuals, though, we had different answers about how willing we were to break out of that paradigm. I think the parallel to how that applies to comedy and culture is obvious enough to leave it to you to run with it.

Comedy, though, often comes with a significant language barrier which gives undue emphasis on the differences between cultural comedy styles. So while Americans and British are said to have different comedy styles, nobody talks as if they are exclusive in any way, just that people from either culture might slightly prefer one over the other. However, people talk about "Japanese comedy" as if it is a completely separate form of comedy.

That is completely untrue, and evidence to the contrary abounds. I have my personal experience in performing comedy to Japanese audiences, but there are also more mainstream examples. The US sitcom *Friends* did very well in Japan, for example. And some Japanese shows have found audiences in the west, such as the obstacle course game show *Takeshi's Castle*.

Culture and comedy have a relationship, just as culture and cuisine, culture and music, and culture and many other things have relationships. Just as people can enjoy other culture's foods because ultimately we all eat, and people can enjoy other culture's songs because we all enjoy music, people can enjoy other culture's comedy because all humans enjoy humour. Culture is just a starting point that provides a context in which certain trends are emphasized and others de-emphasized. Culture doesn't create exclusive types of comedy not be offered or understood by individuals who originated in different places, any more than it creates cuisines that are not edible among all humans.

Cultural biases don't go away so easily. When I've performed jokes in Japanese, some dismissed me as being simply unable to provide "Japanese" comedy. Some found me not **funny**, and didn't care why. Some didn't object to the style of comedy I did, but nor did they enjoy me in particular. And, thankfully, some laughed. However, when you think about it, that exact same range of reactions exists when I perform in English to English speakers. Looked at that way, culture can have significant influence, but is merely one aspect of all the context that goes into creating any one instance of laughter.

Chapter Five

TIMING

5. Timing

☺ Are We There Yet?

> *"Observe due measure, for right timing is in all things the most important factor."*
>
> ~ *Hesiod*

Every comedian knows that timing is *the* single most important factor in getting someone to laugh. Exactly how it works, though, is a little harder to explain. What makes timing in a comedy sense so hard to pin down is that it is not measurable in any sense. It's not like a 100 metre dash where simply being faster is better. Comedians are often said to have a quick wit, but the craft of comedy is not about getting things out there as fast as possible. It's about getting ideas and punchlines out at the "right" time, and how do you objectively measure what's "right"?

It might be tough to measure, but with the right model of how the brain works, we can at least start to see why going too slow or too fast can lead to problems. It's all about managing that all important flow of activity through synaptic connections. Too slow, and the flow of activity trickles through connections bit by bit, never coming together in one moment in a large enough swell that can cross the laughter threshold. Too quick, and the brain literally physically can't keep up with the information being presented, and the flow becomes essentially leaderless, free to go off in different directions that don't lead to areas of critical activity.

That's the bottom line, but "timing" doesn't just refer to that critical moment when the punchline hits and the audience gets it. There's a spectrum of time scales, and each of them have their effect on whether or not potential humour will become verifiably funny. I'm going to divide timing up into three essential time scales, "microtiming", "mesotiming", and "macrotiming". These aren't terms that I've ever heard comedians use, but they will help in getting to a more nuanced understanding of exactly how humour relies so much on getting the timing right.

☺ Microtiming

> *"Time is what keeps everything from happening at once."*
>
> ~ *Ray Cummings*

The smallest level of timing is in relation to that moment when connections between neurons are lit up. As described, when neurons get lit up over and over, the brain responds by strengthening that area with more synapses, which require more terminals on the end of a neuron's axon. For a brain cell to grow new appendages, it takes valuable resources including proteins and sugars. Which also takes a bit of time. Minutes, or even hours, as opposed to the mere fractions of seconds it takes for electrochemical signals to zip about. This process where a neuron devotes the time and energy to growing a new terminal to connect with is called "long term strengthening". Since the actual development of that new tendril takes place over a minimum span of minutes, then there is a lot of play below that time frame in which connections can be activated and reactivated without changing their structure.

Microtiming refers to this scale of time before the brain has a chance to make a structural change in its pathways. It's the current flow of activity, inspired by what is happening right now. It is smaller than any one instance of what we might delineate as potential humour. For example, in terms of joke telling, any comedian can tell you that one missed word, even just a beat or a pause at the wrong time, can make the whole joke fail to land. This makes sense when you consider that the flow of activity necessary to support a funny moment relies on signals travelling within milliseconds between tiny points of connections on microscopic brain cells. How many synaptic connections or cells have to be involved for the humour network to respond is a question we can't really answer. Modern understanding of the brain doesn't give us a model for knowing things like what is the minimum amount of cells needed to be involved to form a single thought process, or how to differentiate one thought from another, or how to even delineate "one thought". However, in terms of humour, if we look at it from the other end, where laughter seems to be sensitive to such miniscule changes in tone or pace or nuance, I think it's reasonable to extrapolate that the physical process in the mind is equally delicate.

We do know the flow of activity in your brain moves at a speed measured in milliseconds, which is a scale much lower than individual words being said by a comedian. At that rate, it's fairly easy for missteps to happen, where the comedian rushes ahead in presenting information, or takes too long to get there. That comedians can mistime jokes is not really a new insight. However, if we consider that all the activity connected to the experience of hearing one joke is connected to all the other patterns of activity before and after, then we can understand more about why timing matters. All the activity in your mind is connected, part of an ever continuing and evolving flow.

One small change upstream can lead to entirely different patterns downstream.

In practical terms, this means that if a comedian makes a misstep early in telling their joke, even if they tell everything afterwards just right, hitting their punchline just as they intended, it might not work. Why? Because the minds of the audience zigged when the comedian zagged, so they weren't on the same path when the comedian landed the joke. The opposite problem can happen, where the comedian timed the telling of the joke wrong, the audience ended up on another path of activity, but it does result in the right conditions for a laugh response, but at something other than what the joke was supposed to be about. This happens all the time when comedians are testing out new material, and in our daily lives. You can be telling something that you mean to be funny for one reason, but in the telling, you say something different from what you intended, and people laugh.

Microtiming is the level at which we are exposed to the delicacy of potential humour's flow in the mind. Any small deviation in the direction of that flow can steer it away from an area with potential for the right synaptic activity.

☺ Mesotiming

"A day without laughter is a day wasted."
~ Charlie Chaplin

In Jerry Seinfeld's documentary movie *Comedian*, at one point, after a bad performance, he laments to Colin Quinn, "I made that rookie mistake of opening with new material." It's fairly common advice thrown around among comedians. Start with something you know is reliable, to get the audience to believe in your ability to be funny, and then you can try out some more experimental stuff. It seems intuitively reasonable, but, why should it matter? Shouldn't a joke that's funny just be funny on its own, regardless of whether or not the joke before it was funny? Any performing comedian can tell you that jokes can influence each other, so the order and success of jokes matters. The reason why is to do with what I'm calling "mesotiming".

"Meso" is not as popular a prefix as "micro" and "macro", so it may not be so familiar. Predictably, though, it means "medium", and it's the level of timing between the very small and the very large. In this context, I'm using "mesotiming" to describe the level of timing that is more than single instances of potential humour and the beats within them. It roughly corresponds to the time a comedian has on stage, or the length of the conversation you have with your friends over drinks. It's a *shared* context of time, in a particular environment.

It should hopefully be obvious that these different levels of timing are reliant on each other and don't have distinct boundaries between them. Whether or not a segue between two jokes is on the level of microtiming or mesotiming is kind of arbitrary, and not worth worrying about. The point is that mesotiming is very much to do with *sequential experience.* Sequence and order are just other ways of thinking about timing, ways of saying now or later based not on a stopwatch, but in relation to other events.

Of course, what makes the order of events important with potential humour is the impact each event has on the context. Let's say you've got a hilarious anecdote about what happened at the doctor's office last week, and you can't wait to tell it at dinner with your friends. During dinner, though, one of your friends tells a sad story about their pet dying. Maybe your anecdote would help bring people's spirits up, or maybe it would seem insensitive to be telling jokes when your friend is down. It depends on the mood. Whether or not now is the time for your anecdote is contingent on the feel of how things are going as the context of the current experience develops, determined by the events leading up to the moment before you decide to start trying to be funny.

That's "mesotiming". The scale on which each experience changes the context for the next experience. How is it manifest in the concept of the flow of activity in the brain? It's a little like microtiming with an added layer. On the microtiming level, it's purely about that flow of electrochemical activity, and how it's zipping around the available pathways in your brain. On the mesotiming level, it's that *plus* the fact that your pathways, the connections between brain cells, are continually changing. The practical difference manifests in the way jokes fail. With a miss in microtiming, potential humour merely fails to inspire laughter, but the audience is still thinking about the topic at hand. With mesotiming, the change in pathways means the audience is now thinking about something different, or at least considering the potential humour from a considerably different perspective.

As your brain cells grope around, connecting and disconnecting from each other all the time, your brain is changing the shape of its neuronal network every moment. Further, what particular positions your brain cells are in, right this very moment, is contingent on what changes they've been making over the last little while. The pace at which it adds up can alter your state of mind, or mood, over the course of minutes, and hours. It's weird to think about, but your brain has a different network now than it did just a few moments ago, making you ever so slightly a different person than you were then. In terms of comedy, what joke you just heard, what your friends were saying just before, what kind of experience you've just had, all changes what kind of audience you are for what comes next.

☺ the Myth Of The Long Set Up

> *"There's never enough time to do all the nothing you want."*
>
> ~ *Bill Watterson*

It's a common mistake among new comedians that they think they *need* more stage time because their jokes won't work unless they set it up just right with all sorts of "necessary" information. Which, given the context of what we know about the brain, might seem like a reasonable strategy. If the brains of the people in the audience are changing their physical networks moment by moment in reflection of the experiences they're having, then couldn't you essentially spend all the time it takes to construct just the right pathways, and then drop a killer punchline that would initiate a flow of activity that rushed through those pathways in just the right way? After all, you helped build them, so couldn't you exploit them exactly as you wanted?

It would be great if it worked like that, but unfortunately it falls apart for one key reason. The way the audience is building the pathways in their heads in response to the set up they are experiencing is not entirely under the control of the comedian. Far from it. Even under ideal circumstances where an audience *wants* to laugh because that's what they paid for, and they love the comedian on stage, and they are interested in where the joke is heading, they can't help but have their own interpretation of everything the comedian says. That influence of their individual interpretation is cumulative. As the inexperienced comedian continues with her long set up, moment to moment she has less and less control over what the audience feels or thinks about what she is saying. By the time the comedian gets to the punchline, almost every single audience member could potentially be in a different mental state from each other, and all of them not where the comedian needs them to be in order to make the right activity happen.

Laughter is a little like a verification of the existence of similar mental pathways between the comedian and the audience, and also, or maybe even especially, amongst the audience members. So long as there is no laughter, there is the potential for any audience member's activity to go off into slightly different pathways, and start creating new pathways, essentially moving further and further away from the agreed topic. An experienced comedian, by keeping the audience laughing, is constantly confirming with the audience that everyone is more or less on the same page. Keeping everyone "close enough" in terms of being on the right pathways needed to inspire laughter.

That divergence of interpretation on its own is enough to derail a joke with too long of a set up. In real life situations, other factors will start to come into play. An audience that is there to see a comedian perform has a certain expectation of how often they will laugh, and if a comedian waffles too long, they'll feel disappointment. That disappointment will factor into their experience, which adjusts their mental pathways, and becomes part of their interpretation. There can be other influences on the audience's impression on top of that, but it's all part of the totality of the mind's experience of events. As a human brain experiences things, it forms opinions about them, and those opinions in turn shape how they then experience things in the next moment.

The divergence between what the inexperienced comedian wants the audience to think and what the audience actually thinks can easily become *too far* for any punchline to succeed. When you consider how fragile a joke can be, where one mis-said word or one pause too short or too long can ruin a joke, it's easy to appreciate how far a comedian can miss if they spend minutes on end working toward some punchlines.

There's an alternate problem, essentially the other side of the same coin, that can happen. A comedian could potentially spend their set up over-explaining a joke with repetitive details, so that by the time the punchline comes, it's not that the audience is all in different places so much as they're all so overly familiar with the topic that they're in a *too close* situation. The most common occurrence mixes a little of both, where parts of the set up that the comedian hammered home are *too close* for the audience, and for other parts the audience mind wandered away *too far*. The pattern of activity in the mind is multi-branched, and very flexible, so anything is possible. The key point is that the audience is free to think how they want to think, and there are lots of ways their mental activity can get away from what a comedian intends.

However, it's not that long set ups are strictly impossible. They're just really, really hard. If, by some mastery of story telling, you can get the audience to follow along with you in just the right way, you could potentially tee up all the right pathways with the potential to be activated in one big splash that makes the humour network practically explode. Hopefully you can see, though, just how difficult that would be. If anyone can pull off a long set up, it's going to be the comedians who have years of experience to guide them. It's definitely been my experience that new comedians have shown me over and over again how long set ups are more likely to get away from you.

🙂 Before It Gets Old

> *"Always leave them wanting more."*
>
> ~ PT Barnum

One time, a long time ago in the days when people rented VHS tapes, my father, his then girlfriend, and I, rented out two Robin Williams standup specials, *An Evening With Robin Williams*, and *Robin Williams: At The Met*. I don't remember which we watched first, but I remember we laughed a lot and had a great time. We put on the second one and were ready for more laughs, but... we didn't. Not as much. There were a few similar jokes, but even with the stuff that was completely different, somehow it felt like we were watching the exact same show over again.

Robin Williams can't be held responsible for staying on longer than he should have, but video tape technology allowed us as an audience to put his performance in a context of breaking one of the golden rules of comedy. Always leave them wanting more. Getting off stage just before the audience has had too much of you is one of the most tried and true pieces of advice in comedy.

Time and time again, though, comedians, even truly experienced and talented ones, break this rule and suffer the consequences. It's a totally understandable trap. When you're on stage and the whole audience is laughing more and more as you go on, it feels incredible. What performer wouldn't want to ride that wave of success as far as they can take it? After all, the whole reason they got on stage is to get that feeling.

This is a good place to point out that the mechanics of how humour works as described in this book is not the be all and end all within the interaction between audience and comedian. Processing humour, and all the thinking that happens in and around it, like all thinking, takes energy. A significant reason why you should always leave your audience wanting more is because they simply get tired.

Still, in addition to the mundane aspects of not wearing out an audience, there is also a component that is specific to the process of humour. The reason we can't just call it a matter of audience energy and be done with it is because there is no consistent model for how much energy it takes to wear an audience out. An audience might watch a half hour sitcom, and feel it was just right. The same audience might watch a comedian do a five minute set and feel the performer was on too long. Equally possible, the same audience could watch a two hour comedy movie and wish it went on longer. If it were simply a matter of brain energy, how could we account for audiences being hungry for more in longer situations and tapped out of laughter in shorter situations? When we put on that second Robin Williams special, we were asking for more, yet still managed to be disappointed by our own reaction.

If you become familiar enough with the performance, you can start to recognize and anticipate the directions it is taking you. Within that framework, it gets harder for the performer to inspire the right activity in the minds of the audience. A comedian can wear out the audience in fifteen minutes, or even one minute, if that comedian presents themselves in a way that allows the audience to develop a familiarity with their overarching approach. This kind of "familiarity" is not necessarily an intellectual understanding that would enable an audience to reliably predict the next joke. Just like you can feel a musician's style is predictable without any ability to create music, an audience that starts to get a feel for a comedian's style, or a sitcom's premise, or a movie's themes, is not necessarily being educated about the techniques necessary to create their own similar jokes. None of us watching those Robin Williams specials could hope to tell jokes like Robin Williams can, but that didn't stop us from feeling that we knew it all too well. It's just that the familiarity an audience gets for a comedian makes any joke they tell start to feel *too close*, even if the details of any one particular joke contains new information.

☺ Macrotiming

> *"Tragedy plus time equals comedy."*
>
> ~ *Steve Allen*

A couple years after the tsunami that devastated Japan in March of 2011, a friend of mine organized a tour of Brazilian comedians to come to Japan to perform for the Brazilian community, which is pretty big in Japan. One of the comedians had to pull out of the tour because he had made some jokes about the tsunami that triggered a lot of hostility against him. People were so upset about his tsunami material that there were even death threats against him. They were, at least in part, hurt and angry that someone could make light of something that had affected them so deeply.

Could that change? As the day of the disaster fades into the past, will it become safe for that Brazilian comedian to come to Japan with his material about the tsunami? It's commonly accepted that with time, not only will it become okay to laugh at tragedies, it's almost inevitable. Which makes sense given that ultimately, as explained, anything and everything can be funny. However, tragic events seem to come with the caveat that it seems to take some amount of time for them to become material to work with. Which raises the question, why the delay? What *exactly* changes to convert tragedy into comedy?

If, for a moment, we can stand to be cold and objective about it, a new tragedy in someone's life is just another set of new information that inspires new patterns in a person's brain. If it's a very significant tragedy it may represent a huge paradigm shift in a person's world view. That shift is manifest in your mind by the development of whole new connections and pathways. So right away, part of the problem in making jokes about brand new information is the tenuous connection it has to the rest of the network. Although new pathways come with new weak connections, it takes some time for all the potential to develop. It has to settle in at least a little, to develop a minimum amount of connections that can be activated to a degree sufficient to rise into the laughter zone.

A tragic event has an added difficulty. No pathways in your mind are completely devoid of connections to other pathways. Every pattern of neuronal connections starts out as a branch of existing patterns. In the case of tragedies, connections will start out initially connected to negative connotations. Feelings of sadness, and anger, and hurtful memories. A comedian trying to make a joke about something that prevails in people's minds as having very few connections, and those few connections are only to largely negative feelings, will have a hard time constructing a joke that inspires novel activity within existing patterns. Possible, but really, really hard. As one example of succeeding earlier in the curve, the satirical newspaper and website, *The Onion*, succeeded in making jokes about the terrorist attack in New York on September 11[th,] 2001, within weeks of it happening, in their September 26th issue, largely by tapping into people's anger and outrage.

Over time, the pattern of pathways in the brain that relate to the upsetting information, no matter how tragic the content it represents is, will get assimilated. Time heals all wounds. At least, it heals all wounds for most people, the ones who don't suffer some kind of permanent psychosis or scarring. People who are permanently scarred by a tragedy are unlikely to ever find it funny, which makes perfect sense given what we know about the mind. People for whom something will never be funny are the ones who pull the tragic events deep into the "trunk" of their tree of mental patterns. They build such strong connections to the hurtful information that no joke can ever overcome the *too close* problem.

This is a more concrete way of describing the process by which someone gets "used" to something. Every time they think about it, and each personal and environmental context that exists at the moment the memory passes through their mind, and each piece of new information that they receive about it, fills out the connections between their interpretation of the tragic event and the rest of their world view. As the patterns in relation to the tragic event deepen and gain more associations, they become ripe for even more neuronal patterns to be made. As more patterns are made, more weak connections become available for activation.

Knowing when any one tragedy will cross over from upsetting to potentially funny is impossible to chart, as there are uncountable devils in the details of any one tragic event. It's an issue of macrotiming, a skill that comedians need to master just as much as the delicate nuances of timing the beats and pauses in the way they speak.

☺ Say No More

> *"Time changes everything except something within us which is always surprised by change."*
>
> ~ Thomas Hardy

From my childhood through to my teens, I loved *Monty Python*. I had all their albums, and listened to them over and over again. I knew all the words to all their sketches and songs. I took pride in the depth and obscurity of my knowledge. For example, I knew by heart the subtle differences between the Parrot Sketch as performed on the TV show and the version performed at *The Secret Policeman's Biggest Ball*. I was one of those annoying people who would shoehorn *Monty Python* bits into conversations, thinking it made me funny. Eventually, I both smartened up about how not funny it is to be regurgitating other people's material, and I also completely burned out on *Monty Python*. I had overexposed myself to the point where you might think that I had made every piece of *Monty Python* material *too close* to ever be funny again.

Decades later, I stumbled across *Monty Python* videos on the internet and watched their best sketches and songs. I still remembered all the lines, and could recite them in anticipation of their key moments in each sketch. It was like returning to the neighbourhood I grew up in after decades away, and remembering where each friend's house used to be. There was every indication that the material would still be way *too close*.

And yet I laughed. A lot. How?

Given almost any amount of time, the same potential humour can be funny again. What's happening is that while the content of a particular joke could potentially be remembered well enough that we could look at it as a fixed constant, it's everything else in the mind of the audience that is around that information which is changing. Within the time frame of one comedy performance, the structure of the pathways in your brain can potentially shift just enough to make more areas of weak synapses available, allowing for something to be called up again and laughed at. Over longer time periods, between performances, that constant change of physical structure in your brain keeps increasing the possibility that you could find the same joke funny again. Not definitely, of course, as it's contingent on how you relate to that joke, not just the time apart from it.

I still had all the information about the *Monty Python* sketches buried in my head somewhere, but all of the context around it, all those patterns of neurons that make me who I am, had changed in the meantime. In other words, forty-something me is, in a very tangible way, a whole new audience for the sketches that twenty-something me is tired of.

Not that time necessarily resets potential humour that you laughed at before to a state of being definitely funny again. You might forget or remember potential humour to any degree, your connection to that potential humour might alter to varying degrees, and so your response to that potential humour will vary as a result. The quantity and quality of your laughter can go up and down in no predetermined way. Nonetheless, within all the possibilities, the *potential* exists for you to laugh as much or more at something you've laughed at before.

Some of the standups I have performed with I've known for over a decade and I have seen some of their material over and over again. And over again. And then again. Some of their jokes I can imitate down to the intonation of every syllable. And yet still, I can laugh when I see it again. Different show, different crowd, different mood, and, importantly, different me. I'll laugh one show, not the next, be kind of amused the next time, then I'm completely unimpressed, and then it strikes me as funnier than ever before, and then it changes the next time, and on it goes.

Of course, as common experience reveals to us, *most* of the time an audience will not be inclined to laugh at the same jokes again if they are exposed to the same potential humour a second time. However, it's not the mere fact that they remember the experience that kills the chances of laughter, it's that there is not enough that is different in the relationship between them and the potential humour to significantly shift their connection to it. Also, in the context of comedy as a craft, audiences generally have an expectation of new jokes from a professional comedian, and their disappointment at not having that expectation met pulls the context around the joke in a negative direction. Which means that, unfortunately, for a performing comedian, you can't rely on the potential for an audience to change the pathways in their brain to be sufficient to keep your jokes going from one performance to the next. It does happen though. People can remember any kind of potential humour perfectly from the last time they experienced it and they could laugh. It's their connection to the joke that matters.

☺ Timeless Jokes

> *There are 'quips and quillets' which seem actual conundrums, but yet are none. Of such is this: 'Why does a chicken cross the street?['] Are you 'out of town?' Do you 'give it up?' Well, then: 'Because it wants to get on the other side!'*
>
> ~ First known written version of the "chicken crossing the road joke", from The Knickerbocker Magazine, 1847, author unknown

Jokes can be written down on paper, and anyone can either pick it up to read it for their own amusement, or act the part of the comedian and tell that joke to someone else. This ability to isolate a joke from its source is one reason why it seems that a joke must be an individual entity.

Go online and purchase a video of Louis CK doing standup, and you are receiving his jokes, but Louis CK isn't personally involved anymore. You can still see him telling you the joke, so it feels like he is still involved, but it's only different from a joke written down in that the video and audio experience is more compelling to us than words on a paper. He's already removed from the joke, but this is just the first step as his jokes seem to get away from him and take on a life of their own. When you remember a joke from Louis CK's set and later tell it to your friends, you're probably going to say "I heard this great joke by Louis CK…" However, if they then pass it on to their friends, after enough time, eventually the "Louis CK has this one about…"

preamble to every retelling will fade away. When it hits that point, has the communication from Louis CK become irrelevant, and the joke is now its own separate entity?

The reality is that the joke was never its own entity, even when Louis CK first told it. Say you and I are both holding our own digital calculators. I type in 2 plus 2, and I see the number 4 on my screen. If I tell you to press 2 plus 2 on your calculator, you see a 4 as well. But did the number 4 somehow jump from calculator to calculator? Was it *my* 4 that I gave to you? Following my instructions, you were able to recreate the same pattern that I had. Yet no separate individual entity defined as *the number 4* actually did any physical moving between calculators. And now that you know the equation, you can tell someone else so they can see the number 4 on their calculator.

The number 4 that you see on the calculator screen is the result of a pattern. A pattern that is made both from the sequence of events of typing in the equation that results in 4, and the physical parts that are the little sections of liquid crystal that can be used to make all the numbers from 0 to 9. It's like how potential humour involves a sequence of events for sensory input, like seeing or hearing something, and then the physical component is your brain assembling the perceptions into something meaningful. A joke is no more a separate entity being handed from one person to another any more than the number 4 is being passed from calculator to calculator.

We live in a culture now that for a number of reasons connects extra information to things like jokes. We attach credit, for example. So, for us, knowing that Louis CK told the joke is part of an overall social contract so that he can get paid for the use of the joke, claim expertise over its proper telling, and establish credibility so that we believe his next joke will also be funny, making us confident that it's worth hiring him. The major difference between hearing a joke from an original source and reading it on paper is the benefit of richer context in which to experience the activity in your mind. While you're listening to Louis CK, you could be thinking "Hey! It's Louis CK telling this joke! That dude is funny!" Thus involving more pathways which bring in more synapses. The way he tells the joke will, I believe, be much more in tune with the spirit behind the joke. He thought of it after all, so the beats, the timing, everything will be tightly bound together to convey the meaning as best as Louis CK intended it. His particular accent and gestures, his exact choice of words, everything that defines him as a person is connected to how he thought of that joke, and is conveyed in how he tells it, which are additional factors for inspiring patterns in your mind. In short, a joke by Louis CK probably sounds best coming from Louis CK.

A joke written down or retold, though, while still potentially funny, is the artifact of a meme that is now removed from all those other possible contexts. You have just the words to work with, nothing more. Which, of course, can be enough. I've read things that have made me laugh out loud, and I'm sure you have too. But whether a joke is funnier when told by the original comedian than when it's read, is determined by a ton of variables within the context of you receiving that joke. All those variables are context that surround the core principle. A joke inspires activity in the mind of an audience, and without the participation of the audience mind, there is no joke.

A joke written down is just an asynchronous transmission of that joke. Only when the connection between performer and audience is completed does it exist as a joke. Until then, it's just words. Like a rose is just a plant until someone receives it as a gift, as described in chapter one. This is more and more a common occurrence in our world as our technology allows for people to communicate across time and space. The nature of communication, and how much we feel that the source, Louis CK for example, is still speaking to us even after the joke has passed through a hundred different people at a hundred different parties, is a question of philosophy that I'll leave to the Wittgensteins and Chomskys of the world. It really doesn't change the fact that every joke ultimately has an original source, and that it is not funny until it lands with someone that laughs, no matter how far the journey or how many people got something from it along the way. The fact that jokes can be written or recorded in any way, says a lot more about our constantly evolving technologies and their ability to facilitate new kinds of communication than it does about humour.

☺ Another Term For "Not Funny Anymore"

> *"There comes a time when suddenly you realize that laughter is something you remember and that you were the one laughing."*
>
> ~ *Marlene Dietrich*

If you're into classical music, when you go shopping for it, there is a large selection for you to choose from. All sorts of different performances of music that could easily have been written over two hundred years ago. Or maybe you're into "world music", which may involve variations in traditional musical styles that could go back as far as humans have ever made music. There are plenty of fans of various musical traditions from different times, and there's no reason to suppose that a tune written a thousand years ago should affect you any less than a tune written yesterday.

However, in book stores, you won't find the humour section to contain much that isn't from recent years. Writing by Jonathan Swift or Mark Twain is over in the literature section, meaning that it's no longer expected to be laugh-out-loud funny as a book by Dave Barry or David Sedaris. They no longer sell you *A Modest Proposal* for its comedy value, they sell you its historical value. The marketplace tells us that some art forms, like music, are timeless, but others, like comedy, have a limited shelf life.

Why should that be? Allowing for a little difficulty in overcoming stylistic or linguistic barriers, such as breaking through the archaic style of Shakespeare that frustrates high school students in English class, certain themes have the potential to always appeal. All people everywhere, and throughout time, are born, die, experience fear and love, and other commonalities of the human experience. Not everybody likes Shakespeare, but enough do to keep him popular. Those that do can really feel something when reading his plays or watching them performed. They can even laugh at the jokes. Let's face it though. Outside of a small subculture of enthusiasts, most people are unmoved by jokes told in Shakespearan English. And even less so if we go back further, to someone like Chaucer for example.

Focusing, then, on the humour part only, jokes from other times may or may not make you laugh, depending entirely on your relationship with them. If you like slapstick, an old Buster Keaton film might make you laugh. Why not? When I was about twelve, I thought Abbot and Costello were hilarious. Now... not so much. When I was young the arbitrary zaniness of a Bugs Bunny cartoon made me laugh. Now sometimes I'll watch a cartoon and I'll laugh less at the arbitrary craziness, but I'll catch more subtle allusions intended for an adult audience that were invisible to me as a child. My cousin Vanessa tells me that her son, when he was eleven, was a huge fan of Laurel and Hardy, and because of the added context of being in the presence of her son's enthusiasm, she could enjoy the movies too. She saw Laurel and Hardy through her son's eyes, which is a whole different meta context for activity to flow around in her mind.

Jokes aren't entities unto themselves, as described before, so they can't contain any permanent or timeless content. It's just a happy circumstance when a joke from another era and a person's state of mind line up well enough to inspire laughter. The difficulty comedy has in reaching across generations is that the chances of a joke inspiring laughter become slimmer and slimmer the more divergence there is between the norms of the era the joke was made and the norms of the era when the joke is experienced. Music seems to tap into a different process in the brain, something more visceral that is less contingent on how the performer and the audience relate. Comedy simply doesn't have that luxury.

Sometimes I've seen old movies or videos of old standup performances marketed as "classic comedy". For comedy, "classic" is less of an assertion of quality and more of an excuse for the lack of impact. Personally, any time I see the term "classic comedy", I know I'm not going to laugh. Well, I might laugh a little. But usually, I just find jokes from other eras to be quaint windows into other worlds where people found different things funny. I think comedians like Buster Keaton, Charlie Chaplin, Laurel and Hardy, or whoever else from whatever era, deserve respect for tapping into the zeitgeist they lived in and successfully drawing comedy from it. As much respect as any comedian working today. It's just that, somewhat sadly, respect doesn't make me laugh.

Chapter Six

THE COMEDIAN

6. The Comedian

☺ No Comedian Is An Island

> *"Remember, we're all in this alone."*
>
> ~ Lily Tomlin

In 2001, a German named Armin Meiwes decided he wanted to try eating human flesh, so he put out an ad on the internet asking if there were any volunteers to be eaten. Amazingly, a guy named Jurgen Brandes stepped up, so they got together, and cannibal dude ate edible dude. The German police heard about it, and Meiwes was arrested in 2003. Despite the victim's willingness, Meiwes was convicted of manslaughter in 2004, which is when it made headlines internationally. I have a video from that same year where I, like a lot of comedians at the time, riffed on how bizarre it was that there was a guy who wanted to get eaten. My bit concluded by saying that the story gave me hope, because surely if two people with such depraved and unique desires can find their match, so can I.

Does that premise sound familiar? It might, actually, because the exact same joke was used in the pilot episode of a TV show called *Committed*, which aired in 2005. It was a short lived mediocre sitcom about a man and a woman in New York who have a budding relationship. In the pilot episode, on their first date, the male lead tells the female lead the story about the German cannibal, ending up with the exact same message of hope that I ended with. The similarity was close enough that as I watched the scene play out, I felt a very eerie sense of *deja vu*. Did my joke in some happenstance way get passed around by word of mouth until it ended up with some writer for sitcoms in Los Angeles? It's not impossible, I suppose, since memes do get around. That's not what I think really happened, though. The fact is that it's just not that brilliantly original of an idea.

It would seem that the incalculable variability between human brains would support the notion that different comedians would necessarily create completely different jokes. With all the potential difference, the odds of any two people, let alone two comedians, thinking the same thing about the same topic might seem near impossible.

However, in developing the patterns that make up our mind and our thoughts, we are not islands. We are partly shaped in a general sense by the environment we find ourselves in. You are always taking in cues from the people around you about what the group and other individuals are doing, and what they are feeling. They are doing the same as well. In this environment, we are all networking our thoughts and feelings, so that as we associate in groups, we build consensus about what is happening. After all, the ability for us to find some kind of workable consensus so that we can operate within groups is a cornerstone of our success as a species.

A comedian then, has to be sensitive to this shared information. Making a funny comment is not about throwing out an opinion or new information to the audience. Being funny is about being aware of the current status of the shared information held by the community, and being at play with the flow of thoughts and perceptions in and around that information.

The "group" can be many things, and often it's a matter of layers that overlap. For example, a comedian getting on stage to prepare for a set of standup comedy has to cope with many levels of group dynamics within the audience and what information they share. Are you performing at a corporate event where everyone works together, or a comedy club where there are separate groups of friends, and many individuals that don't know each other? The comedian has to be aware of the group in the context of that particular room. Is this a show at a theatre where people paid an admission and are expecting to laugh? Or is it at a bar where some of the patrons may have just come in for a drink and had no idea there would be a comedy show? Is it a younger crowd that will get that one joke about a new trend? Is it an older crowd that will likely know about an old reference?

A lot of these considerations come to comedians intuitively, more commonly referred to as "talent". It would be a recipe for failure for a comedian to try and intellectually work out all the many levels from general society and human nature all the way down to the individual people sitting in the front row that night. Even if most comedians are riding a wave of intuition about these various levels, nonetheless, to look at it objectively, what they are doing is being sensitive to what is the relevant information flowing through the community of their audience, however that community is defined.

Groups of people, and the consensus between them about the information they share, is pretty much by definition more general and less varied than the perceptions of individuals within those groups. Since comedians are working on a level of relating to group minds, then we should not be surprised to see that different comedians come up with similar material. It's not so much that two individual comedians happen to think alike, it's that the group they are performing to does.

Every comedian at one time or another sees another comedian do a joke about the same thing in the same way, and one of those comedians, usually the less established one, has to abandon it. It's inevitable, really, because the whole point of developing material is to draw upon shared patterns of thought in collective minds, and select from the razor thin margin that exists in between ideas that are *too close* or *too far*. From that point of view, I'm not so bothered that I saw a joke of mine simultaneously created elsewhere. It indicates I was at least playing to my audience, although clearly I needed to work harder to get out of the *too close* zone.

What I'm more bothered about is that my ideas overlapped with a sitcom that was not very popular and got quickly cancelled. I would be a lot happier if I had thought of joke that was similar to something on *The Office*, which is the show that replaced *Committed* in the same time slot. It would be a little more encouraging to have something in common with the writers of shows that go on to critical and popular success.

☺ What If Richard Pryor Was From The Suburbs?

> *"If you ain't funny then get the fuck off the stage, it's that simple."*
>
> ~ Richard Pryor

There are quite a few unfortunate stereotypes surrounding comedy and comedians. Things like, "women don't make as good comedians as men," or, "people from minorities, like black people, make good comedians because the hardships their culture has experienced gives them a different perspective," or, "Jewish people are just naturally funny." While stereotypes can even seem flattering, like the last one there about Jewish people being funny, in the long run they are detrimental because they shape our expectations, mostly erroneously, toward biases that limit our capacity to consider more accurate truths. For every good stereotype that exists, there is an unsaid opposite corollary, which can be just as harmful as a negative stereotype. Saying one group of people is funnier is saying the others aren't as funny. In other words, stereotypes are always harmful generalities in one way or another.

Really, I'd much prefer to just dismiss this topic altogether, because it's the kind of area where the process of even acknowledging some arguments lends them more credibility than they deserve. The short answer is no, there isn't any evidence of any biological basis for differences in the ability to either laugh or tell jokes between any group of humans based on categories of race, ethnicity, or phenotype. "Race" isn't even a meaningful differentiator at all in the realm of science, but that's a whole other book.

"But wait," I hear you say, because you're so fucking hard to convince all the time. "It's not that black people in the US are funnier because they are born black. It's because they are raised in a culture that is outside the mainstream white culture, or in social and economic conditions that give them a certain perspective," or something like that. "In terms of comedy, race isn't a factor on a literal, biological, level. It's a short hand for social conditions that often coincide with racial issues. In other words, wasn't Richard Pryor funny because he was from the ghetto?"

No.

Lots of unfunny people come from the ghetto, too. And every other social economic class as well. One thing we have to take into account is that humans are consistently terrible at factoring in all the examples of when something didn't happen. It's called a "spotlight bias". We tend to only think about the stuff we see in the spotlight of our focus, and totally ignore everything in the surrounding darkness, where there are probably countless counter-examples. If you count all the "black" comedians in the United States, or even in the world, who have achieved any degree of fame, you are talking about mere handfuls of people. Statistically speaking, compared to the larger group that would identify as "black", it's nowhere near enough to indicate any particular propensity for comedy. And that's not even taking into account the distortions of selection that happen because of marketplace considerations. Isn't it pretty much inevitable that *Saturday Night Live* will have at least one "black" cast member in every season? Does that represent the reality of the statistical distribution of funny people among various ethnicities in the United States? Or does it represent a television producer's sensibilities in trying to match the American audience's current expectations about diversity?

If you look into your own life and think about people you know personally that make you laugh, your sample will always be infinitesimally smaller than what is available in the wider world. Whatever categories you use, if you are saying that "All X people are naturally funny", you are selecting the people you know who are, and ignoring all the people you don't know who aren't. And no matter how popular you think you are, no matter how much you think you stay on top of pop culture, the second group of people who you don't know will always be way, way larger than the group of people you do know. The difference between the two is so huge as to make the conclusions we draw from our personal experience absolutely irrelevant to the reality.

We can dismiss a lot of stereotypes based on the fact that humans are just really bad at figuring out trends and generalities based on anecdotal observation. When we really look at the numbers, most assumptions are based on unfounded generalities that are invisible from a statistical perspective. Which would be true enough if there was an even playing field, but there isn't. Which comedians are presented to us for consideration are as much a matter of opportunity, marketing, and media distortion as is the case with just about any other performance art.

Still... Since the potential for inspiring patterns within that shared world view is based on the construct of individual brains, and individual brains are partly influenced by the culture they grow up in, then isn't it possible for that construction to take effect collectively? Couldn't a world view that can be mined for comedic material be shared within a culture, enabling most of its members to be funny?

Yeah... maybe. Just not supported by much, if anything, in the way of evidence. Take the stereotype that Jewish people are funny. Not funny amongst themselves, because that's not particularly special. Every group of humans based around any level of commonality uses humour amongst themselves to bond and consider themselves to share jokes that others can't fully appreciate. The real question is if Jewish people, as a result of their culture, are more disposed to being funny to people in other cultures. Could that be possible?

First we have to separate out the concept of subjective appeal from objective appeal. Just like no individual comedian is simply funny to everyone, it's unlikely a culture would be funny to every other culture. Japanese audiences, for example, don't seem to have any particular appreciation for what we might identify as "Jewish comedy", even if translated. The TV show *Seinfeld* , is arguably a product of "Jewish comedy" as both head writers, Jerry Seinfeld and Larry David, as well as a significant portion of the cast, are of Jewish heritage. The show was a massive hit in English speaking culture, but when made available in Japan as *Tonari No Seinfeld* (*My Neighbour Seinfeld*) it got just about zero response. Not merely because comedy doesn't survive translation. As a counter example, the US sitcom *Friends* was very popular in Japan. Which doesn't disprove the potential for one culture to be funny to another, it just means that Jewish culture has its own audience. It's not Japan, but maybe it's the English speaking world.

The stereotype of Jewish people being funny does seem to be supported by a large representation of Jewish comedians. Exact numbers are hard to come by, not least because defining who is a "Jewish Comedian" isn't exactly cut and dried. David Cross actively denies being Jewish, but some people claim that decision is outside of his control because he was born from a Jewish vagina, as he puts it. Whatever the number is, the stereotype exists, and so does the possibility.

It could be the case that Jewish culture values being funny, and the desire to meet that expectation encourages people within that culture to pursue funniness. In other words, it's not that a culture can endow people with any particular ability, but it can endow them with the drive. Sure, why not? That doesn't necessarily mean they'll succeed, though, and there are significant obstacles to a culture promoting funniness. For one, a culture can't do much to foster any particular approach to being funny, because any tradition or formula that could be passed from one human to another can become familiar and made *too close* in the mind of an audience. As awesome as he was at standup, does the world need *another* Woody Allen on stage? Another problem is that trying to maintain group cohesion by relating to each other, while at the same time fostering an ability to appeal to people outside the group, are conflicting goals at worst, and a very tricky balance at best.

The possibility exists, but is the proportion of people of Jewish heritage working in comedy in the United States evidence of this potential having come to fruition? Unfortunately, the current state of the world doesn't do anything to let us know. It could be possible that many cultures have potential to make others laugh. It could be that a high percentage of aboriginal Australians are wickedly funny, or would be if they lived close enough to other cultures to practise against them, but they live in the outback of Australia where there isn't a lot of mass media and cultural mixing. Ultimately, we can't make a fair assessment of cultural groups being more or less funny because the world is the way it is, different people are where they are, and there's no way to construct a comparison that is separate from the context in which we do the comparing.

What is definitely clear is that *biologically* there is no evidence of potential funniness being based in any way on genetic heritage. As for culture, it could hypothetically be possible, but without massive changes in the opportunity afforded to communities of people around the world, there is no objective evidence to support what seems to essentially be just another stereotype. A stereotype that only hurts the people it supposedly compliments. Jerry Seinfeld, Woody Allen, Joan Rivers, Robert Klein... All the comedians born into Jewish culture, like any comedian from any culture, had to earn their reputation. To suggest anything was handed to them just by being born into it is to diminish their efforts.

☺ Women

> *"There is not one female comic who was beautiful as a little girl."*
>
> ~ Joan Rivers

In the January 2007 issue of *Vanity Fair*, there was an article by atheist activist Christopher Hitchens titled *Why Women Aren't Funny*, which attempted to articulate a common bias that women are not as capable as men at generating humour. Hitchens' essential premise is that men need to be funny in order to attract women, as a sort of mating dance. For women to reverse that function and try to appeal to men is essentially as unnatural as a female peacock trying to display her plain feathers, and for that reason it doesn't really work.

The mating dance theory starts to look really suspicious if we just scratch a little under the surface. For one, it's an ethnocentric question. In Japan, for example, there isn't much bias against women in terms of comedic ability, which is saying a lot in a country that otherwise has many issues with gender equality, especially in the workplace. Not that sexism doesn't creep into the comedy world in Japan at all, but the mating dance theory doesn't seem to be a component.

In any case, when I get on stage and start telling jokes, the men and women in the audience laugh equally. No men seem to be very threatened or laughing less. Much to my disappointment, women don't seem to be swooning or laughing more. This is very unlike any kind of mating ritual we see in other species. Male peacocks, for instance, don't respond one little bit to other male peacock's tail feathers. If one male peacock has really awesome feathers, the other males don't gather around to compliment him in any way, instead, they'd rather he was dead. Other male peacocks with better feathers are just competition, pure and simple.

It's almost certain not to be a mating dance, but the stereotype that women might not be able to generate comedy as well as men is harder to waive off than other stereotypes, like those based on ethnicities or cultures, because there is some evidence to support the idea that human brains do have material differences based on gender. One difference, for example, is that male brains appear to have more cells devoted to processing, referred to as "grey matter", and female brains have more cells devoted to networking between cells, referred to as "white matter". It's clear that men and women both have the ability to process humour, but tangible differences in brain structure could imply differences in exactly how that processing occurs. Differences that may possibly lead to observable differences in levels of ability to appreciate or generate humour.

One study at Stanford, for example, suggested women and men are no better or worse than each other at getting jokes, but they do have slightly different expectations about how funny the jokes should be, which in turn impacts their appreciation. Men want the joke to be funnier, so they tend not to laugh quite as much because it turns out the punchline isn't as good as they hoped. Women, on the other hand, don't have expectations as high, so they find it easier to laugh because it's easier to exceed their expectations. The difference in expectations, however, is so slight that you can only really see it in a highly controlled lab environment. Which is also true for most, if not all, studies on gender difference and humour. I sure as hell can't see much difference in a comedy club, where individual variability completely obscures any general trends. Peacocks would have died out if their mating displays were that subtle.

What about making comedy? What accounts for the under representation of women in the professional comedy world? As far as I've observed, comedy troupes are almost always more men than women. From *Monty Python* in the 60s and 70s to *Kids In The Hall* in the 90s and 00s, which were all men. For both groups, scenes involving female characters often had men playing the parts in drag, which then became part of a larger meta-contextual joke. Personally, I know from running comedy shows and trying to bring together teams of improvisers or comedic actors for various projects that getting funny women is almost always the hardest part. In the years that I've been with the group of standups that I regularly perform with, we've only ever had one or two women on our roster at any one time. Sometimes we've had none. When it comes to scene based comedy, like for a sitcom or something, female actors with comedic timing can be found, but it's harder to find women who can help write, develop, and perform material.

While there are definitely a lot of very funny women, they do seem to be a minority. Is this evidence to suggest that women on average aren't as funny as men? Are the women who are funny merely imitating the traits of men on some level? What about our anecdotal experiences with cognitive differences? Women won't read maps and men won't ask for directions. Men are from Mars and woman are from Venus and all that?

Are little girls rewarded the same way as little boys are for being the class clown? Which comes first, girls not being appreciated for being funny because they're not as good at it, or girls never trying to become funny because no one appreciates it? I don't know, but that's my point. No one does. There's just too much distortion from society as a whole to make categorical claims about a gender's influence on potential comedy talent. There is no research that gives us any reason to believe there is anything in our brains that would explain why, for example, over the history of *Saturday Night Live*, from 1975 to 2008, only around 30% of the cast have been women, and most of those women were members in the later years when it almost certainly became a mandate to increase their representation. There are expectations on women, and men, in society that make it hard, if not impossible, to discern how men and women would act if societal expectations of gender differences did not exist.

Women celebrities tend to follow an archetype of young, attractive, and sexually available. In another article in *Vanity Fair*, written by Alessandra Stanley in response to Hitchens' article, called *Who Says Women Aren't Funny*, the accompanying photographs of female comedians Tina Fey, Maya Rudolph, and Kristen Wiig, depict them in an unmistakably sexy way, in an effort to make the point that sexy and funny are not exclusive qualities. The degree to which sexiness and funniness can or can't be mixed is an interesting topic, but the more important point is that it's not a topic men have to contend with as a matter of course the way women do. For example, Amy Sedaris, in the comedy TV series *Strangers With Candy*, went out of her way to make her character on the show as unattractive as possible, effectively addressing the sexiness issue by trying to subvert it. Whether embracing it or denying it, it seems a woman who wants to succeed as a comedian has to orient her persona around her sexuality. You can be hot or not, but it's tough to be neither. Men, on the other hand, aren't automatically expected to have any particular disposition to their sexiness.

In the social realm, a lot of times when I tell people that I perform comedy, people assume that it means I have an easier time talking to women. However, stage-funny and friend-funny are different, and no women, or men, want to be performed at in a social setting. The ability to break the ice with a little humour can help, but at a certain point one has to get off stage and be genuine. At that point, it gets harder to mask individual incompatibilities with comedic talent.

Many studies show that both women and men use humour to evaluate potential mates, but in different ways. Women seem to want men to demonstrate an ability to be funny, and men look for women who laugh at their jokes. However, that's not necessarily the end of the story, because it might not be that humour is a cause of attraction, maybe it's merely a corollary to other desirable traits. Similar to how the type of woman who might be attracted to men with visible displays of wealth, like a nice car or expensive watch, doesn't necessarily care about cars or watches, but use those as an outward indicator of underlying resources. A man who can make a woman laugh might also be a man the woman can get along with. Similarly, a woman who responds to a man's particular brand of humour is also conveying that she thinks like him. Ultimately, it's the empathy that's desirable, not the wit. Also, it needs to be noted that some studies indicate that women might laugh at bad jokes by men they like and be unamused by jokes that get good responses in other contexts when told by men they don't find appealing. Indicating that, if you're a man, when a woman laughs, she thinks you're funny because she likes you already, not the other way around.

None of which says that women are necessarily any worse at being funny just because of the different ways men and women use it to test potential mates. Especially in the context of performance comedy, which is a whole different craft from social settings. Even if it were true that there was possibly some small mating dance element to humour, there's no reason to believe that how comedy is used when forming relationships has any bearing on how one succeeds on stage or in any comedic profession. The standup comedy world has no lack of aspiring comedians who have been told by their friends that they are so funny that they should be on stage, only to be consistently met by stone cold silence when speaking to an audience of strangers. The ability to speak to large groups of people and appeal to their aggregate sensibilities is completely different from an individual man seducing an individual woman. So different that there isn't any evidence to suggest that either the individual woman or man would have any better or worse time crossing over into the ritualized world of comedy performance.

As a closing note on the topic of women in comedy, consider George Burns and Gracie Allen, a comedy duo that started out in the vaudeville days, and progressed into the early years of film. At first when they performed together, they assumed the roles of George Burns as the funny man and Gracie Allen as the straight man. Because, of course, women are pretty and men are funny. However, despite the situation being set up to match societal expectations, which were even more stark in the US in the 1930s, the audience found Gracie funnier, and the duo found success when they picked up on the audience's perception, and switched roles. I like that anecdote because it defies the perception that any woman who performs comedy is fighting against her natural tendencies to be the recipient, not the creator, of comedy. Were that true, there should have been no need for Gracie and George to switch roles. She couldn't help but be the funny one, though. It was just who she was.

☺ The Outsider

"I never told a joke in my life."

~ *Andy Kaufman*

Andy Kaufman, after he became famous for playing the character Latka Gravas on the successful sitcom *Taxi*, would refuse to do the character for audiences that came to see him live. Instead, he would read from the book *The Great Gatsby*. At first, the audience would think it was the set up for a joke, but he just kept reading. In some performances, after the audience reacted by booing, he would put on a record, which was a recording of him reading the same book, and starting right where he left off. As the audience came to realize that he was seriously going to just read, and there was nothing else to the performance, a lot became upset and left. Hearing about it after the fact as an anecdote, I find the concept hilarious. Had I actually paid to see that performance, though, I might feel like my time and money was wasted. Same joke, but the funniness is determined entirely by the point of view of the audience.

If humour is a matter of just the right amount of connection, not *too far* and not *too close*, doesn't that mean that people who are too weird, too different, will by definition not succeed? It's definitely a bigger challenge to make a deliberately unconventional act appeal to a humour network that is set up to appreciate a certain level of familiarity. However, it's actually not relevant at all how weird the content of a comedian's act might be. What matters is if the comedian can establish a relationship to you, and then they can channel pretty much anything through that connection.

It was never the case that Andy Kaufman was coming at the issue of comedy by being purely random and seeing what hit. He was famous for working as a busboy at a restaurant even when he had achieved fame, in order to keep himself real. Though he also joked(?) that he kept the busboy job just in case his comedy career didn't pan out. In any case, as weird as he was, he made an effort to stay in touch with the sensibilities of "regular" people. Even though his ideas were coming from the side of *too far,* he knew how far to reach in order to bring them to the funny zone where his audience could appreciate it. Maybe he came by it naturally, maybe he did it deliberately, and he had misses as well as hits. If he hadn't had the ability to make hits consistently enough, though, he would have been ignored by the public and we wouldn't be talking about him now.

If you're an outsider who can't make the connection to your audience, you remain merely an outsider. Bridge the gap and bring your insane ideas to where the audience can get it, and you have a shot at being funny.

☺ A Computer That Can Be Programmed To Tell Jokes

> *"One day ladies will take their computers for walks in the park and tell each other 'My little computer said such a funny thing this morning!'"*
>
> ~ Alan Turing

Heather Knight, a social robotics researcher at Carnegie Mellon, has been programming a robot to perform standup comedy. As far as performing comedy goes, on one level, she's got the right approach. She's actually researching how to make computers interface better, with standup as a test environment, and so her focus has been on making the robot respond appropriately to crowd reactions or information present in the environment in order to select what to say next. Unfortunately, that next step is a doozy. At present, the algorithms that drive the performance can only select from a predetermined list of options which are all written into the program by humans. Some human standups use material written by others, so it's closer to a real standup than you might think. However, it's still a long way from a computer that can generate material.

For a computer to be able to get beyond emulations and construct an original joke of its own, it would need to establish a relationship with its audience. Humans, as a result of our evolution, are in a sense "designed" to come together in spite of the potential for our immeasurably complicated brains to go off in different directions. A computer would in some way have to share that capacity for finding compatibility with people, to participate in our communities. Or at least be able to simulate participation, in order to be at play with the mental patterns the humans have. The computer doesn't have to have the same world view, but it does have to have an insight into the human existence well enough to bridge the gap between it and us. The same way a really surreal and strange human comedian succeeds in spite of being very different from the rest of us by knowing enough about how the audience feels in order to bridge the gap.

I don't think there's any way of programming straight algorithms into a computer that will make humans laugh. If any one human could quantify some style of humour down into algorithms that could be programmed into a computer, then any other human listening to the resulting jokes could perceive the formulae in use, and could get too used to them to continually be amused. On the other hand, it might be possible to make a computer that can learn, in whatever time scale, enough about us that it can relate to us.

Looked at that way, humour isn't the challenge with computing, relationships are. Can a computer be developed that can look at the world in a way that is similar to us? Can it *relate* to us? Maybe, only time will tell. When you consider how humans from two different cultures laugh at different things to the point where we almost think they have different concepts of humour entirely, you can see that computers have a potentially very large gap to overcome in order to relate to us.

More interestingly, to me at least, is that computers manufactured together and created for similar purposes could be considered part of the same "community". That being the case, I suspect that artificially intelligent computers are very likely to be telling jokes that are funny to each other, and maybe not so much to the humans listening in. In the end, I'm not so worried that computers will be funnier than me, just that they'll be laughing at me behind my back.

☺ Holy Ground

> *"It's always funny until someone gets hurt. Then it's just hilarious."*
>
> ~ Bill Hicks

Michael Richards, famous for his role as Kramer on the sitcom *Seinfeld*, was doing a standup performance in 2006, and got into an argument with some audience members. During the altercation, he shouted some racist stuff at them. The incident was caught on video and passed around the internet. If he weren't previously famous, the incident would be forgotten, if it were even noticed at all. Beginning comedians at open mike nights the world over routinely say obnoxiously unfunny offensive things, and no one really cares enough to make it an issue, they just don't laugh.

Other comedians have had it worse than Richards, though. For example, in 2013 a human rights court in Vancouver awarded 15,000 Canadian dollars to Lorna Pardy, to be paid by comedian Guy Earle, for hate speech he allegedly said to her in an argument that started during a standup comedy performance of his in 2007. The case was a complete miscarriage of justice, bringing a human rights tribunal into what was essentially a childish spat in a room full of people who, according to court records, were all too drunk at the time to give reliable testimony.

Predictably, I side with the comedian in that I can't see how anything said was genuine hate speech, but that's because I disagree over the definition of hate speech as it was applied in that situation. Something that is an ethical argument outside of the scope of this book. What does relate to this book is another disagreement I have, which is with a lot of the comedy community about the sanctity of the stage.

A lot of comedians, from beginners on up to the most established performers, express a belief that the stage is a place that should provide them with a special freedom of speech. It doesn't, not in any legal sense, but also not in any performance sense. Which at first might seem to disagree with the premise of a humour network that is sensitive to all aspects of potential humour. If jokes told by a professional comedian on stage is a different context than jokes told off stage by your friend, doesn't that imply that a default assumption built into that context is the audience knowing that you being on stage definitively means everything you say is a joke? If that were the case, you could potentially go down any road, touch any topic, say it in any way and direct it at any person, and the audience will know you're merely being ironic because you're on a comedy stage.

It's possible that some audiences will be that forgiving, but I would argue that it's optimistic of comedians to assume that audiences will default to thinking, "Oh, I didn't like what he just said, but he's performing comedy, so I know he didn't mean it." In my experience, audiences tend more to think, "Why is this guy on stage if he can't come up with anything more clever to say than *that*?" They came to a comedy show, and maybe paid some money to get in, because they are entering into an agreement that the whole reason why the show organizers put a performer on stage is because that person has demonstrated an ability to say things in a funnier way than people in other situations. If a comedian fails to live up to that promise, then the context of being on stage not only won't save them, it could even work against a comedian because the audience expectations come down from a higher place.

When people get angry and hurl verbal insults at each other, their desire to say something hurtful can motivate them to reach for convenient epithets. Especially when arguing with someone not known very well personally. Whether or not an insult based on broad stereotypes reveals an underlying bigotry or just an attempt to be hurtful by any means available, is a debate I can't settle. All I can say is that in any case, audiences expect comedians to handle verbal banter better than civilians on the street. A guy shouting from his car at the person he thinks cut him off in traffic may lack the verbal tools to rise above a bluntly prejudicial statement. The lack of tools doesn't make the statement any more forgivable, but it does make it unsurprising. A comedian, though, loses some credibility when they can't do better.

When a comedian is confronted with a truly obnoxious heckler, they need to be able to rise above the casual bigotry of throwing around words like "dyke", and instead reach for more nuanced descriptors, such as someone being an obviously overly entitled and delicately sensitive hipster. Just as an example.

Comedians desire a space in which they can be free to experiment, which is what the stage at a comedy show provides. However, that experiment shouldn't be testing what an audience will put up with, it should be testing what the comedian can produce. Fail the experiment, and there is nothing stopping you from being exposed to libel suits, criminal charges, or social persecution that exists for anyone in any other situation. Making a defence based on a notion of special permission will fail, because there are no extra privileges afforded to comedy, not by society, and, far more importantly, not by the audience.

You're not a comedian just because you're on stage. You should be on stage because you're a comedian.

🙂 Irony

> *"Satire is a sort of glass, wherein beholders do generally discover everybody's face but their own."*
>
> ~ *Jonathan Swift*

Mark Smith was tasked with booking a comedian to perform at the White House Correspondents' Association dinner in 2006. By his own account, he hadn't seen much of Stephen Colbert's work, so he probably didn't anticipate the direction it would go. What ended up happening was one of the most talked about comedy performances ever. With President George W Bush, arguably the most powerful man on planet Earth at the time, just an arm's length away, Colbert satirized the president under the guise of praising him, in the ironic style that has become Colbert's trademark character. Colbert's performance dramatically polarized the audience into two camps. On the one hand were those who felt that it was masterful irony, hysterical comedy, and a truly courageous moment of speaking truth to power. Like the *New York* magazine, who described the performance as "brilliant". On the other hand were those who felt that Colbert had bombed, and there was nothing to his act but the uninformed rantings of a political malcontent. For example, a columnist for the *Washington Post* referred to the performance as merely "lame and insulting". Was the dividing line between those who get irony and those who don't? Those who understand humour and those who don't? Was it merely a difference of political views?

Some people think of irony as meaning some sort of cosmic justice, where having done something wrong leads to a retribution that fits the crime. Others define it as a sort of elevated sarcasm. Irony can also be used to describe being self aware, as in wearing clothes you know are outdated, but doing so "ironically", as in, you know the clothes are out of fashion, and anyone looking at you should get the idea that you are doing it deliberately because you're an annoying hipster.

In sarcasm, one makes it obvious that they mean the opposite of what they are saying, through intonation or manner. With irony, one does not try to make it easy for the listener by giving clear hints about the fact that what is being said or done is only a joke. This is the kind of irony I will be focusing on, the kind of irony that was the driver behind Colbert's performance for President Bush. The act of telling a joke as if you were being completely serious, without much, maybe not any, clue for the audience that one is joking. It is up to the audience to figure out for themselves that a joke is being told. In some cases, the irony can have even more impact if there is someone who does not understand the comedian is being ironic, making the irony even funnier for the audience observing the interaction. For example, everyone loves a story about how someone misunderstood an article from the online satirical news site *The Onion* to be a real news story, such as when China's *People's Daily Online* mistook an article about North Korean leader Kim Jong Un being the "sexiest man alive for 2012" as being real.

    Irony overlaps not only with sarcasm, but also with satire. To me, the difference, if there is any, is that satire leans more toward pretending to be a source other than yourself. For example, the old British puppet show *Spitting Image,* which ran from 1984 to 1996, satirized politicians by creating puppet caricatures of them, and speaking their lines as if the politicians being mocked were the source. Irony, however, is when you are not trying to be anyone else, but are being yourself, saying things that people will hopefully realize aren't actually your views. Satire is saying, "imagine if this other guy said this thing." Irony is saying, "it's me saying this, but you know what I *really* mean." Though, the dividing line is blurry. Stephen Colbert isn't pretending to be a specific other person, but he is playing a role of an archetype, pretending that archetype is him. Sarcasm weaves in and out of irony as well. Personally, I take sarcasm and satire to be tools to use *within* irony, but if you prefer to think of them as being separate but related or overlapping concepts, that's your call. I don't think it will make much impact on my definition of irony that I'm using here, which is that irony is the act of telling a joke while simultaneously pretending that it's not a joke, with varying degrees of how much you give hints to the audience.

If you, or any of your friends, like to use ironic humour, you have probably seen for yourself in many social situations how this precarious balance between joking and sincerity can sometimes fail, and fail spectacularly. When a regular joke about an offensive topic fails, an audience often feels that the comedian was merely ham fisted, mishandling an attempt to be funny. Failed irony, though, because it is not clearly delineated as a joke, has the potential to leave the audience convinced that the comedian was being sincere. They're not merely failed comedians, they're people with messed up world views.

I often experience foreigners living in Japan stating that Japanese culture has no irony. The evidence being that attempted ironic jokes fall flat with Japanese audiences. However, it's interesting to note that everyone who fails to convey an ironic joke assumes a lack of perception in the audience, and very rarely, if ever, considers their failure to convey the whole context. Irony is tough even in your own culture, so when trying to convey it across cultures, and especially across languages, it shouldn't be surprising that given you don't know as much as your audience does about the best way to convey what you want to convey, that what you're trying to say might not come out as intended. It could definitely be argued that irony is appreciated to different degrees in different cultures. However, the idea that some cultures simply don't have it, as if some humans were simply incapable of grasping a form of communication that other humans grasp, is, I'm sorry to say, a mild bigotry.

In Japan, there is a lot of emphasis in the culture on consensus, which is hard to do if you have to be always monitoring for insincerity and double meanings. I have been told many times by Japanese and non-Japanese alike that the Japanese people regard sarcasm as "mean" and irony is just not really used. Which is simply not true. There are Japanese comedians, Hitoshi Matsumoto is probably the most known example, who provide a good amount of irony, and has audiences that love him for it. His film Dai Nipponjin, released as Big Man Japan in English, is entirely ironic, depicting the comedic failings of an underwhelming super hero as a serious documentary. From my experience, most Japanese aren't prevented from appreciating irony and sarcasm because of some internalized cultural resistance, it's just not given a lot of focus in popular culture due to restrictions imposed by the conditions within broadcast media. In any case, the Japanese do create and enjoy irony, although arguably in different amounts than other cultures. More to the point, the mannerisms and contextual clues that let people know that irony is taking place can be just different enough to make anyone trying to convey it across a cultural divide reconsider what we take to be obvious.

Even if a native speaker of English tries to convey irony to another native English speaker, where both of them have the same cultural background, there is still no guarantee of success. The act of trying to pretend to be sincere as much as possible while at the same time letting people know that it's only a joke, is definitively a struggle between which impression is going to make the most impact. Confusion over whether or not irony was successfully conveyed will always be a risk, regardless of language, culture, or context.

In terms of how the humour network perceives it, irony is all about having that extra context that the audience brings to the joke. When two people hear some kind of irony, but only one gets it, we could argue that both people heard and saw the exact same information. The person who laughed though, at some level understood that this information was not simply for receiving, but to interpret as a joke. In other words, asking themselves the question, "what is the comedian *really* talking about?" The answer to that question is a whole *additional* flow of activity, on top of the information strictly contained in what the comedian offered. That additional activity that the audience developed on their own by answering the question is the one that provides an even wider context, a broader pattern of activity, that inspires more synaptic connections.

Critically, that additional layer of mental activity represents the audience member's own understanding of the topic at hand. So, in a way, irony is a lot like spontaneously thinking of something funny on your own. In other forms of humour, the comedian provides most of the insight that leads the way for synaptic connections to be made. In irony, it is left much more for the audience to try and flesh out a broader context with less to go on.

Think of the challenge that raises for the comedian. They have to be sensitive enough to the audience's world view to inspire patterns that the audience can make themselves, by communicating just enough to make that happen, and not too much to make it obvious that is what they are doing. Irony is a tough form of comedy.

Does this take on irony imply one has to be smarter to get it? It seems that to get an ironic joke, the audience has to have more information than the comedian is providing. However, consider that the additional mental activity used by the audience to support the joke the comedian is offering can be anything. It can be as simple as a feeling, a shared sentiment between the audience and comedian that they don't like the same politician or something like that. In other words, for those who take a little pride in thinking that irony is an indicator that they're just a little more clever than those who can't get irony, sorry, but that's not the case. Getting irony is, just like all comedy, an indicator that you are of a similar mindset to the person being ironic, nothing more.

😊 Zaniness

> *"Any community that gets its laughs by pretending to be idiots will eventually be flooded by actual idiots who believe that they're in good company."*
>
> ~ René Descartes

In the realm of comedy in Japan, right now, zaniness reigns supreme. Japanese comedians often wear wacky costumes and have outrageous personalities, usually accompanied by catchphrases and hooks that they will use over and over again, what they call an *ippatsu gagu*, or "one hit gag". For example, a couple years back there was a dude named Katsura Sando, whose act consisted of counting out loud, and every time he said a three or a number divisible by three, he would say it in a strange voice and manner. That was his whole act. No, really, that was it. Other examples include Kojima Yoshio who basically just went around saying, "it doesn't fucking matter!" to anything and everything while wearing nothing but skimpy Speedos, and Masaki Sumitani who called himself "Hard Gay" and dressed like a bad stereotype. What I've told you here is pretty much all they had on offer.

Zaniness is the opposite of irony. Completely removing any subtlety and shouting, "I am being funny!" Personally, I hate that approach to being funny, but, people laugh, and why not? For some people, being spoon fed the indicators that comedy is happening simply creates different starting places for the activity in their mind, and for them they can then progress on towards areas where maybe they can get the right kind of activity for activating the humour network. For me, obvious indicators of zaniness makes me feel like the comedian is trying to force the terms of our relationship, and so my brain goes down pathways completely unhelpful for that comedian.

That's why comedy is ultimately democratizing. By separating out content from the process of how humour works as stimulation in the brain, we can see that zaniness is not inherently more or less funny than anything else, it's just another context, another pattern of mental activity. Some people will like it, some won't.

I stand by my hatred of zaniness, because doing something like wearing a big zany bow tie is not much different than starting out the comedy set by declaring, "I am a *comedian*, so things I say are therefore *funnier*". I don't want to be told you're funny, just be *actually* funny and let me decide. But, conforming to my own proposal of how humour works, I have to admit, zaniness is not by any objective measure a worse way of doing comedy. It's just not for me.

No one type of comedy is right or wrong. If they laugh, it's funny. However, that statement doesn't stand on its own without considering who "they" are. "They" are the audience, and they create the conditions under which comedy can succeed as much as any comedian does.

Chapter Seven

The Audience

7. The Audience

🙂 The Stepford Audience

> *"I think comedy as an art involves the audience as a participant as much as it involves the artist."*
>
> ~ *Craig Ferguson*

Audiences at the beginning of a comedy show are a random set of individuals with different reactions. As the show progresses, bit by bit they all gravitate to one place. By the end of the show, they're all laughing, or they're all quiet, or they're all leaving, or whatever.

I don't think it's particularly new to suggest that the reason the audience all comes together into one feeling is because, over the course of a show, they feed off each other's reactions. People doing comedy shows around the world know this, and sometimes turn this to their advantage by planting someone in the audience who will laugh a little extra in hopes of nudging the audience toward a favourable mood. In Japan, these people are called "*sakura*", which literally translates as "cherry blossom", and the implied meaning is that these audience members bloom a little earlier than the others.

The effect of one person laughing is the message that it is okay to laugh. It releases inhibitions about what is acceptable to laugh at, because we all monitor the degree to which we laugh to keep it in line with what we think is socially acceptable. If one bold person can establish ahead of us that it's okay to let it out a bit more, then we're inclined to follow suit. Similarly, people who are noticeably agitated or disdainful can make people feel a little more uptight. It's not an effect that is limited to comedy, it's all social situations, and I think the tendency to head toward consensus of some kind can be observed in many types of group situation.

Throughout all of this consensus building within a group, there is a very important element that is particular to comedy. If we take this group mentality as just being the result of achieving a comfortable consensus between audience members, then then we can see why they might all be a little happier, or bored, or whatever. But does it explain why they all seem to find the same things funny or not?

Being in a good mood built upon being in the presence of happy people is one thing, but *getting* a joke is another. To actually get a joke requires that I have certain pathways in my mind that potentially allow for the right kind of activity to happen. And in the context of a group, we all need to have similar patterns in order to get similar activations. It doesn't seem so obvious that a good mood shared by people around me will make me *get* a joke more than I would otherwise. Good or bad mood, my ability to understand would seem to be roughly the same. I might be more open to *laughing* if I'm in a good mood, but that's a different thing from getting the joke in the first place.

As your brain is figuring out which pathways to strengthen and select for continued use, one of the indicators of a "good" neuronal pattern to keep is that everyone around you agrees it's a good pathway to have. Collective relevance is a big thing for humans, since we have evolved to be socially interdependent. It's good to be on the same page with other people, to know what others know, and to be assured that what you know is true in the eyes of the group. When audience members hear each other laugh, it means that not only is the joke personally relevant, but collectively relevant. When you hear someone laugh at the same joke as you, it raises the context from "*I* find it funny" to "it *is* funny", where factual reality is determined by the consensus of the group. That perception of factual reality is very compelling to your brain, so you favour the pathways that represent it, steering your brain into a different direction than it would have had you heard the joke when you were alone.

The audience comes to a consensus that the material represents a community experience. If half the room laughs, and half the room doesn't, whether or not you do will rest a lot on your individual assessment of the joke. But as more people in the room laugh, you might be more inclined to think there's something to it. You can, of course, stand apart from the group with enough determination to do so, if, for example, your sense of self identity rests on thinking people around you are dumber than you are. There are never any absolutes when talking about the brain, just tendencies, and the tendency for humans is to want to be a part of a group.

That critical pattern necessary for getting a joke, then, is shaped as a group, forming not only a consensus in how we react, but a socially networked pattern that represents our sphere of expectation. "Sphere of expectation" is actually a term used a lot in improv...

😊 Improv

> *"Every time you go the way the audience expects, they'll think you're original. People laugh with pleasure at the obvious."*
>
> ~ *Keith Johnstone*

Keith Johnstone, in his book *Impro For Storytellers*, relates an anecdote that is often passed around in improv circles. He asked a member of the Boston Improvisation group why they bother getting suggestions for scenes from the audience. The improviser responded that it ensures the audience knows that the scenes are actually improvised, and not planned. Just at that moment, an elderly man from the audience interrupts and asks, "Excuse me, but how much do you pay the people who shout out the suggestions?"

When improvisors on stage are so in tune that they seem to know what the other person on stage is going to do before it happens, it seems like it *must* have been planned in advance. Ironically, one of the highest forms of praise an improviser can get is when an audience member says, "Bullshit! That was all scripted!" Improvisers, on the other hand, know that planning ahead actually tends to ruin a performance. Good improv is about having such trust and communication with other improvisers that you get a consensus going in the moment, where no one person is driving. Ideas seem to come from the group as a whole. In that environment, trying to insert a premeditated idea seems rigid and unnatural, and both audience and performers alike can feel the presence of something that doesn't belong.

The audience is as much a part of an improv performance as the improvisers. It's not just that the audience offers suggestions as starting points for scenes. That comes more from individuals within the audience anyway, and some improv groups do away with that formality altogether. Improvisers are in continual contact with the group mind of the audience, using their laughter, silence, and other responses to inform almost every decision that moves improvised scenes forward. The sphere of expectation, in a sense, is a collective agreement on what pathways are available to everyone in the room, performers and audience alike.

The "sphere of expectation" has lots of particular meanings and interpretations to improvisers, but for our purposes a basic understanding will suffice. Essentially, if you have improvisers performing a short scene about two teenage boys fighting over who gets to go out with the girl they both like, suddenly having a Martian step in and threaten to destroy the planet will seem arbitrary and most likely a lame attempt at trying to be funny by being wacky. If the improvisers are acting in a scene about astronauts on a space station, then the appearance of a Martian won't feel so out of place. For each scene, there exists a whole set of potential ideas that could be introduced into the scene, but have to be close enough so as not to seem like worlds colliding. This cloud of potential related ideas is the sphere of expectation.

The sphere of expectation is not absolute, it's relative to each scene. It would be just as weird for the father of one of the astronauts to suddenly step into the space station and warn one of the astronauts to not come home past curfew as it is for the Martian to be in the scene about the dating teenagers. The sphere of expectation is not static, either. It flows and evolves as each new idea is introduced. It might be lame if a Martian suddenly showed up in the scene about dating teenagers when all we know is that they are two boys interested in the same girl. But if small details were slipped out, bit by bit, maybe they could build a context in which the Martian is not a concept *too far*. They both go to Roswell High. One is a transfer student with an unclear past. Girls have been disappearing... Then one of the two boys reveals he is a Martian who has been abducting girls because... Mars needs women!

In other words, you can get to just about anywhere from anywhere in a scene, as long as you push the envelope of the sphere of expectation without jumping outside of it. As my improviser friend Chris says, "you can take the train to crazy town, just don't take the express." In fact, it's critical for good comedy that the sphere of expectation be constantly changing, because the funny ideas are the ones that are just on the boundaries. Right on the edge between acceptable for the scene but still new and interesting.

You can easily see for yourself how this relates very directly with the *too close* and *too far* model of humour in the mind. Revealing the Martian angle in the wrong context is *too far*. Not doing anything unusual and just having teenagers talk is *too close*. If timed right, though, the moment when something out of the ordinary happens, called "the tilt" in sketch comedy and improv circles, such as a teenager revealing that he is a Martian, can be really funny.

The expansion of the sphere of expectation is not determined entirely by the performers. Far from it. The improvisers are paying attention to the laughter, or lack of it, to know what is *too far*, what is *too close*, and what is on the edge of the sphere. It's the same as when two audience members laughing at the same joke inspire within each other a context that says the joke is universally funny and not merely personally funny. Improvisers are exploiting the mechanics of that interaction to write a scene in the moment as determined not by what any one of them on stage thinks is funny, but what the group, the whole room, both performers and audience, collectively determine what's funny.

This is part of how ideas in a good improv scene seem to come from almost nowhere, plucked out of the ether. The resource is the shared mental pathways of the entire room, and if improvisers give in to it, they seem to have the same ideas at the same time, happening as if by design, leading some to conclude they must have scripted it.

There is a lot to doing improv well, of course, that is built upon how improvisers work together as a team and trust each other, following certain principles and developing skills related to creating stories and acting them out. Improv is not merely some kind of Zen act of giving into a nebulous group mentality. The people on stage should be there for a reason, that reason being that they have practised and trained their skills so that when they draw upon ideas from the audience and the group mentality, they know what to do with it.

☺ Please, Have Another Drink

> *"I went out with a guy who once told me I didn't need to drink to make myself more fun to be around. I told him, I'm drinking so that you're more fun to be around."*
>
> ~ *Chelsea Handler*

One of the few times I have come close to being in a fight in my adult life was when I made fun of a drunk patron at a British-style pub I was performing at. During my act, he was loudly hitting on women at the table right in front of the stage, obviously too drunk to really be aware that there was a show happening and that he was being disruptive. I threw a couple of verbal put downs his way, only to get him to shut up. He loudly called me "cunt" until some other audience members shuffled him away, and after the show he threatened to beat me up, and might have carried that out that threat were it not for the presence of other friends bigger and more capable of fighting than me.

Alcohol has diminishing returns in loosening up an audience. Any comedian can tell you that an audience that has been drinking is generally easier to work with than a sober audience. It's just that it has to be under the threshold where they start calling you "cunt" and wanting to kill you. Assuming we stay under that threshold, does alcohol make things funner? Yes and no. No, because among the short term effects, in addition to the loss of motor control and focus, alcohol does damage to dendrites, the branches that extend from neurons, which impedes electrical and chemical signals. However, under the right conditions, one effect from alcohol can become a significant player in comedy, the reduction of inhibition.

Laughter is almost *always* suppressed, at least a little. As social creatures, without even having to think about it consciously, we're always taking care that our outward behaviours are acceptable in terms of the social rules we have grown up with. We don't scratch, pick at, or touch ourselves in ways that people around us would find gross. It's not just potentially icky behaviour that is constantly being managed. Tone of voice, gestures, posture, and just about anything else that makes an impression on others, all reflect our inner set of rules for acceptable behaviour. This suppression of certain behaviours is so well practised by the time we're adults that we usually don't have to consciously monitor it.

Laughter is, by definition, out loud and the point of it is for others to hear it. Since it's something that is intended for others to hear, it naturally falls into the category of activities that we keep in check. A badly timed laugh can upset someone else or embarrass yourself, so there's always a part of you asking, "is it okay to laugh?" It's not a binary function of laughing or not laughing, you just hold back a little, adjusting the tone, intensity, length, and volume as you see fit. If others start to laugh, then you get the message that it's acceptable, and you might loosen up a little and laugh a little more. Alcohol short circuits that process a little bit. It helps you take the cap off your normal suppression of laughter. What it doesn't do, though, is anything to assist the processes that lead up to the humour network responding. If anything, the effect alcohol has on brain cells probably inhibits a person's ability to fully process any potential humour.

The evidence in comedy clubs around the world, though, is that the lack of inhibition more than makes up for any other effects alcohol has on the brain which might work against humour occurring. Of course, just because you're a little less inhibited doesn't mean you're necessarily going to laugh more. If you don't find things funny, your brain can end up anywhere. If you end up somewhere where you're not that happy, and you're drunk, you'll express that a little more than you would sober. This is why hecklers also tend to be drunk. A heckler, if they're a bitter drunk, can start to bring a room down, and reverse that cycle of laughter inspiring laughter.

Other drugs, like marijuana, might have a similar effect of lowering inhibitions, increasing laughter, and then, in a social setting, possibly leading to more laughter. What about nitrous oxide, commonly known as laughing gas? It's actually not as effective at making people laugh as they make it seem in movies and television. It's used as an anaesthetic, and can cause all sorts of effects, like euphoria, dizziness, loss of sense of self... in other words, it gets you high, like any other drug. Then you laugh at things because you're high, assuming those things are funny because the laughing gas made you see it that way. How a drug affects you has a lot to do with your expectations and state of mind, so since there is a cultural expectation of laughter from laughing gas, people almost certainly precondition themselves to exaggerate the laughter response. However, consider that nitrous oxide has been used by dentists as an anaesthetic for over a hundred years. Would they really use it if it made their patients uncontrollably jerk and convulse their mouths in laughter, when the mouth is precisely the area dentists need to work on? Like any drug, laughing gas *can* make you laugh, but so can a lot of stuff that is equally, if not more effective, so it's not really impressive enough to live up to its name.

As you get into more mind altering substances, like LSD, they may actually also have significant impact on how activity flows around the brain, and maybe even on how the humour network responds. However, until they start serving other drugs besides alcohol in comedy clubs, I'm content to limit my analysis to the kind of drugs that are served in bottles, with their pushers being legal corporations.

🙂 Hecklers

"There have been some very extreme hecklers in audiences whose bile was so hateful and so mean that it would be a bit frightening to think that all I'm doing is jokes and yet someone hates me that much."

~ Jo Brand

One time at a show in a small venue that I was performing at, there was heckler who stood near the stage, loudly mumbling, "yeah yeah yeah," in a flat monotone throughout the show. He would say it in response to things comedians were saying, but his timing was strange, and the audience couldn't help but be completely distracted. It totally ruined one comedian's set, to the point where that comedian got off stage early, frustrated and upset. I got on stage with the intent to shut this "yeah, yeah, yeah" guy down, and I pulled him on stage, intending to embarrass him into giving up his game. However, once I could actually see him under the stage lights, it became clear very quickly that he had no idea that I was even interacting with him. I don't know if he was exceptionally drunk, or stoned, or if he had some kind of disability, but he just wasn't all there. There was no way of connecting to him.

This is the worst kind of heckler, someone who is either mentally unstable, or perhaps far too drunk, to the point where there's no human interaction anymore. No comedian can perform with a person like that in the room, and nothing can be done to convince the person to stop. The person doesn't even know that they are being a problem. Eventually, some people just escorted the "yeah, yeah, yeah" guy out of the room.

That's by far the rarest kind of heckler, though. The most common hecklers, you might be surprised to learn, are kind of supportive. Most people's image of a heckler is the kind of drunken loudmouth who shouts "You suck!" when the comedian is having a bad night. Those exist, of course, but more common than that are people who just want to talk back to the comedian, responding to things the comedian says. Sometimes agreeing, sometimes disagreeing, sometimes just saying things they think are important. Which sounds mostly harmless, but they can still cause a comedian to lose their train of thought, or throw off the timing of a joke.

A heckler, as I define it, is anyone in the audience who yells out anything during a comedy set, whether supportive or not. There's a wide variety of intention and manner among hecklers, which makes handling hecklers a delicate craft. If a well intentioned person in the audience yells out something supportive and the comedian throws back a knee-jerk, anti-heckler insult, then that comedian can come across as mean and defensive to the audience, and lose the support of the room. I've seen so many new comedians lose an audience because they are so worried about losing control of their set, that they can't even really hear what a heckler says, and they automatically shout put downs into the microphone to crush the person who dared speak out of turn. The experienced comedian is in the moment, able to evaluate a comment from the audience and respond on the right level.

Of course, people interact in various ways at all sorts of performances, but we can separate out a few other kinds of disruptions at other events to narrow down to what makes a heckler at a comedy show a distinct thing. For instance, you could get someone trying to disrupt a politician's speech by shouting out the opposite political viewpoint, but I think it's rare that during a speech someone who came in as a supporter of a particular politician at some point suddenly turns against that politician so much as to start heckling. More likely a heckler of a politician is someone who came in with something against that politician's policies. A heckler at a comedy show, though, is someone who didn't necessarily intend to come to the show to heckle a particular comedian or a particular type of comedy. I've often performed in bars where a number of patrons didn't even come for the show, they didn't even know there would be one until we started, and yet those people can also end up as aggressive hecklers.

There's also a distinctive aspect to the type of participation hecklers engage in. For example, at rock concerts fans can be so spontaneously moved by what they hear that they jump on stage and dance around for as long as security allows them to get away with it. I was at a concert once where someone was so worked up they grabbed the microphone and started singing. What these fans don't usually do, though, is start performing their own original music, different from what the band is doing. The heckler at a comedy show provides their own material. Whether hostile or supportive, it's not just parroting or following the comedian's lead. They want to express their own take.

Being spontaneous in their disposition to the performer and contributing original content are the factors that, to me, distinguish a heckler at a comedy show from other kinds of audience participation. Hecklers at a comedy show are an extension of something that the whole audience feels on some level, which is that they are a part of a particular kind of relationship, one that emulates the kind of relationship that humour facilitates between friends. The relationship being emulated, that of humorous interactions that exist within social groups, is usually a two way street as far as our brains and humour network is concerned. So it's only natural that some people might slip into creating contributions of a higher order than just laughter response. This is the core cause of comedy hecklers. At some level, they mistake the performance context for being a social context, where their input is as valid as it would be among a group of friends.

Are hecklers an inevitable product of a comedy show? Can they even be a good thing, giving more direct feedback than the subtle cues of the audience as a whole? It's impossible to say hecklers are good or bad as a whole, because heckles vary so wildly. Some people are just jerks, and the kinds of heckling they do will always be unwelcome. Even if we could reduce those jerks, what should we make of the people who try to positively contribute? It is true that when a comedian gets a heckle and makes a snappy come back, that improvised moment can be more precious than any of the predetermined jokes in the comedian's routine. The trouble is that the decision to create such "banter" is almost always made unilaterally on the part of the heckler. Nobody likes being manipulated, comedians included, and the rest of the audience didn't come to watch a show where some random person is dictating the terms. They came to watch a show where they trust, or at least hope, the people on stage are on stage because they've built up credibility. Hecklers break that contract by unilaterally deciding that they are also part of the show, despite the fact that the audience has no reason to believe they are as entertaining as the comedians.

Hecklers aren't likely to go away, since the craft of comedy is derived from a basic human nature that was all about communication between people. Equally so, I'm sure that comedians will always be disdainful of hecklers, because no comedian writes material thinking, "I hope other people say their own stuff while I'm performing this!"

☺ Five Minutes Of Grace

"The crowd, no matter who you are, will give you grace for like five minutes."

~ Colin Quinn

Masatoshi Hamada, a famous Japanese comedian, once remarked that as he got more famous, it got harder for him to know what was really funny and what wasn't. This was because the audiences that came to watch him and his comedic partner Hitoshi Matsumoto, adored them so much they would laugh at anything they said. "I get angry when they laugh at even our bad routines," he said.

The reason a famous comedian can get extra laughs is simple enough to explain in the context of brain activity and pathways. The additional frame of reference that this is a famous person is part of the overall context that the audience is using to perceive the performance. The audience's relationship to this famous personality, their expectations, the history of other funny things this person has done, and whatever else, are all additional pathways for the brain to work with.

If the famous comedian is on form, then the audience will have it confirmed in their mind that they were right about how funny they thought this comedian would be. That concept, the affirmation that the comedian is as funny as they should be, can provide a continued boost throughout a comedian's act. If, however, the famous comedian bombs, then the flow of activity in the audience's mind may move further and further away from territory that helps the comedian. Now they are thinking about how disappointed they are to watch the comedian not be anywhere as good as they expected. They could be even more disappointed than seeing an unknown comedian bomb, because their expectations come further down from a higher place.

One of the many things I like about comedy is that it ultimately levels the playing field for everyone. Fame will buy you a little time, and being unknown forces a comedian to work harder to gain the audience's trust. Eventually, though, both the famous and non-famous comedian will have to contend with the same challenge, which is to build a rapport with the audience. It doesn't really matter about your age, gender, ethnicity, or anything else that we use to divide people into groups. If you make people laugh, you're funny.

Probably a more significant effect being famous has is in preselecting your audience. No matter how big a comedian is, the number of people who are *not* fans will always be bigger than the number of people who *are* fans, but the non-fans don't come to watch comedians they don't like, so as a famous comedian you get to perform for a filtered crowd. The fans not only bring their desire to have their fandom validated, but also a general commonality in their world views to start from, which is probably why they are fans of that comedian in the first place.

Still, all things considered, fame is only one part of a set of circumstances that makes up how the audience initially perceives you. Whoever you are, the moment you get in front of your audience, they begin forming their own perceptions about you, and there's never any guarantee that those perceptions will always be the ones you want them to have.

Chapter Eight

FUNNIER

8. Funnier

☺ Making You Laugh... With Science!

> *"To make comedy, maybe you just have to work hard and be funny."*
>
> ~ *Tina Fey*

Humour, and by extension comedy, is based on a specific set of functions in every individual human brain. However, because the humour function exists to respond to circumstances in relationships between people, there can never be one, simple, surefire way of making any of the people in any relationship laugh, because of the simple fact that no two relationships between humans are ever the same. Even the same relationship between the same people changes over time. It's a constantly moving target, and learning to ride the waves is forever going to be an art. Even the greatest comedians can't give you formulas for what will work, because what works for them might not work for you. Their relationship to the world is different from yours. However, what they can give you are helpful tips for you to reflect on in your journey to be funnier, which is all anyone can do, and what this last chapter is for.

What if you weren't satisfied with that, though? What if you're a super villain, unconstrained by tedious details like "feasibility" or "possibility"? You want the power of making people laugh, guaranteed, every time, by any means necessary. Since we've identified a physical process in the brain for humour, isn't it *theoretically* possible one could essentially force the humour network to fire up every time we wanted? Sure, we can already short circuit past the humour network and directly stimulate a laughter response with tickling, but that's like being handed a pointy stick when you wanted a death ray. The point is whether or not we can take control of someone's *humour network*, forcing them to find things *funny*.

    We have already seen some potential for this. In the operation on a 16 year old girl referenced earlier, scientists were able to tap directly on the humour network and inspire a laughter response. In an effort to resolve the confusion of having the network stimulation without any clear external cause, her brain would look for sources of humour in whatever happened to be present, such as when she was looking at a picture of a horse, in a sort of *post hoc ergo propter hoc* way. With the right super advanced technology, then, maybe we could fire a ray at someone that stimulated that person's humour network with an electrical pulse or something. If our laugh ray worked right, it could be a hell of a laugh, since we wouldn't just be gently tapping one node on the network like the doctors in that operation, we'd be lighting the whole network up like a Christmas tree. Though there's probably a limit on just how much response the laugh centre in your *motor cortex* could generate.

    I don't know about you, but if I were a super villain, and I'm not saying I'm not, I wouldn't be satisfied with that. The person being zapped by the laugh ray in this case isn't going to laugh at what I want them to laugh at, they're going to laugh at anything and everything around them at the moment. If they happen to be looking at a picture of a horse when I zap them, they'll laugh about horses. In the words of every great super villain everywhere, "Bah!"

I don't want to merely build a *laughter* ray that forces people to laugh more effectively than tickling or drugs. I want to build a *comedy* ray that makes people laugh because of concepts I want them to laugh at. To do so, we'd need to force the kind of activity in the brain that the humour network listens for. Which could be possible. In and around all the pathways of the brain, at any one time there are uncountable numbers of weak connections between neurons that, if stimulated, would in turn pass on a wave of signals that the humour network recognizes as something it should respond to. Since this comedy ray would hit all of the potential weak connections at once, it could potentially inspire a huge laugh response by making the humour network think there was something funny going on *everywhere* in the brain.

However, have we really recreated the physical experience of getting a joke? When a comedian gets you to laugh, you are inspired to think on a particular topic, which we could say is like having the flow of activity in your brain directed down a particular path. All the weak connections inspired along the way relate to that topic, so there's a sequential relationship among all the activity in your brain, and the experience makes sense. In comparison with that, you can appreciate how the comedy ray could be completely jarring. We have no way of really knowing how a brain would try and make sense of it, but if all the sudden disparate and completely random parts of your brain were activated without any previous activity to link them, I would speculate that when being hit with the comedy ray, you would first laugh, and then wonder why and think back to nebulous free associations that don't add up to anything concrete. A vague sense of various concepts just out of one's grasp, like a dream that you just woke up from, but have forgotten all the details of, except that it was really, really funny.

Similar to how the laughter ray made people arbitrarily laugh at whatever was in their environment, the comedy ray is making people arbitrarily laugh at whatever concepts their brain structure happens to make available. It's still not making them laugh at the ideas I have that I want to put in their brain, to laugh at *my* jokes. To do that one would have to somehow physically put specific external thoughts into their brain, which on its own is debatable if that is even possible, and then push those thoughts in a direction where they would meet an area of weak synaptic connections, while still keeping the integrity of the new idea and not going into pathways that are too much of the audience's construction ... Yeah, this isn't going to work.

By imagining the mechanics of how to physically force someone to get a joke with science fiction technology, it sheds light on a very real fact about comedy. It's an art, not a science, and will remain so pretty much forever, up until we achieve technologies we can't even imagine today. There are no formulas, no patterns, no mechanisms by which we can short circuit the process of relating that drives how humour is created and experienced.

There is, of course, a way that humans have put thoughts into other people's brains and got them to laugh, and they've done it since before civilization. It's called *being funny*. It's tricky and nowhere near as reliable as a ray that forces physical reactions in the brain, but it can be done, as you have no doubt seen and experienced all the time. So we're going to put away the comedy ray and focus on making people laugh the usual way.

☺ Connoisseurs Of Comedy

> *"I prefer smart audiences because smart people don't heckle. If a smart person doesn't like a comedian, he just blames himself for not having more assiduously researched his entertainment options. Stupid people shout, "You suck." Smart people think, "I suck, for not Googling him."*
>
> ~ *Emo Philips*

For a while at a venue I performed at, there was this group of bankers, at least that's how they identified themselves, who would regularly come to shows and ruin them by talking amongst themselves so loudly the rest of the audience couldn't hear the show. If they ever did pay attention, it was only to heckle and be jerks. One time, as I was standing to the side of the stage while another comedian was performing, I heard one banker talking about the show to another banker, saying, "you've got to be really good to impress us." So, in other words, even though he wasn't listening or paying attention, he nonetheless somehow expected to be entertained, like expecting to find something good on television without turning it on. There's no way of explaining to a guy like that how you can't enjoy a relationship if you don't participate in it. You can only wonder what a crappy boyfriend he must be.

Since comedy is a relationship between audience and comedian, you can't only talk about what people on the giving end are doing, you have to also look at people on the receiving end. An audience unwilling or unable to focus on a performance can not be reached, no matter how talented a performer is. The relationship that makes comedy possible is a mutual creation. Therefor, it's not only possible for people who want to be funny to look at humour and comedy in a way that can help foster a better relationship to their audience, an audience can also influence the relationship in order to enjoy it more.

Even the most willing to laugh audience we could imagine is still more likely to *not* find things funny, at least from the point of view of pure probability. The number of pathways in their head that have the potential to trigger sufficient synaptic responses will always be far, far outnumbered by the number of available neuronal networks that don't.

That said, the relationship between audience and comedian isn't static, so if you increase your odds of laughing a little by following the advice here, then your laughter signals to everyone else, including the comedian, that there is something funny happening, lowering their inhibitions and increasing the context in which they can also find it funny. Everyone else's laughter feeds back to you as well. That helps guide the comedian, contributes to the over all mood, shapes the context of the environment, and validates the content of the joke in terms of community relevance. If all goes well, then the whole situation is set up so that the next joke will have better odds of working than the previous one. So it's worth doing, finding ways to allow yourself to laugh more, because then it means that what started out as a minor adjustment in frame of mind can pay off eventually in terms of having a hell of a good time laughing.

So what do you do to increase your odds of laughing, but without just forcing yourself to laugh in spite of your impressions? I don't think anyone wants to fake laughter, even if it could theoretically pay off in genuine laughter later. People don't go to see a comedian and make the show work for the performer, they expect the performer to offer something that they can receive. Fortunately, there are two things an audience can do to increase their good times that don't involve forcing any deliberate behaviours.

The first is that generally speaking, you up your chances of enjoying a comedy performance more by going with friends. Elevating a joke from individually funny to funny for a group provides a greater context which means more pathways, which means more synapses that could activate the right way at the right time. Although the rest of the audience already contributes to this effect, it's a little more effective with your friends, because their opinion carries more weight with you. Those friends should be enthusiastic or at least open minded, though. The presence of a friend who is really negative can do as much to bring you down as an enthusiastic friend can do to bring you up.

Friends or not, being in a group of people will still do more to help the humour than being alone. Given a choice between watching a video at home or seeing it in a theatre, you are more likely to laugh in the theatre. Following the same logic, I think there's a possibility that live comedy might inspire more humour response than something transmitted by whatever media. Only because the performer adds even more dimension to how people in the room validate the humour beyond the individual level.

Another benefit to involving other people is that it may also help a person to enjoy comedy shows a second time by being with people who are seeing it for the first time. There is nothing stopping you from enjoying the same joke many times over, depending on how much your brain has changed between performances, but the added context of a friend is an additional boost. Even though you know the punchline, seeing your friend get it and laughing will inject a little of that sense of community into the joke that can make you perceive more dimension to it. Although, this has risks. If you are eagerly anticipating your friend to laugh at the same thing you are, and they don't, your disappointment might be more crushing than if you simply didn't enjoy the joke. Further, the more you are focused on anticipating your friend laughing, the more you are thinking about meta-contextual issues outside of the joke, which may take you further away from the right context in order to laugh.

Humour is, at its roots, a social mechanism, so it makes sense that the more social it is, the more effect it will have. Laughter is meant to be heard, so the more you expose yourself to hearing it and having yours heard, you are exploiting its fundamental nature in order to get a better response. Which brings us to the second piece of advice, which is to pick your battles. There is a huge difference between going to a comedy show with a frame of mind of, "I wonder if these guys might be funny," and, "I doubt these guys are funny, but I'll see them anyway just to check if I'm right about that." Just like some people just aren't meant to date, there are some comedians you just won't find funny. All relationships have to start with at least some commonality, or they face some very significant obstacles. When it comes to comedy, some obstacles are probably far too large to overcome in the time frame of a single comedy performance.

Taking an extreme position in any direction will adversely affect the comedy you choose to enjoy. It might seem like it would be beneficial to go to a comedy performance with an attitude of, "I love these performers so I know they will make me laugh!" However, even a great comedian may fail to deliver what you hoped for, and the gap between your experience and your expectation may push your frame of mind toward disappointment.

You're better off not trying to impose too strong a world view on the comedy performance. The comedian is trying to lead you to laughter, to build a bridge between their world view and yours. The less flexible you are on where you stand, the fewer avenues the comedian has to reach you. The tricky part about trying to evaluate your own bias is that you can't know for sure that the comedy on offer and the comedy you hope to see don't match until you actually go see the show. But if you understand that comedy is about relationships, and not a completely objective stimulus like a pin prick, then you have a better chance of evaluating your chances of laughing.

On the other hand, just as a reminder, keep in mind that this is not an absolute rule, just a nudge toward increasing your odds. While laughter is social, with social implications and influences, ultimately your mind is still your mind and it will always come down to you and whether or not you go along with the group, or don't, for whatever reason.

☺ How To Be Funny With Your Friends

> *"Laughter is not at all a bad beginning for a friendship, and it is far the best ending for one."*
> ~ *Oscar Wilde*

If you want to drive a comedian crazy, next time you meet one at a party, ask them to tell you a joke. Comedians hate that. They don't hate it the same way lawyers and doctors hate being asked for free advice. It's because comedians know that their material isn't funny in the social context of a party. Just like being friend-funny doesn't work on stage, being stage-funny doesn't work with friends. Stage-funny is about connecting with a group, a society, a larger context than a handful of friends sharing drinks and hanging out in someone's kitchen. Comedians know that if they try and recreate their stage act in this limited environment, it won't work, and then the people who asked for the joke won't be impressed, and no comedian wants to be seen as not funny. If the comedian doesn't attempt a joke, though, everyone thinks the omission is a lack of confidence or ability, and again the comedian is left looking like they're not funny. Different comedians have different ways of deflecting this, but however they deal with it, it's a situation they want to get out of, not in to.

Not understanding the difference between stage-funny and friend-funny doesn't just cause awkward moments in the social lives of comedians. There's an expectation that humour, not comedy, is a talent that can be applied in lots of situations. People think that if they can be funnier, they can make more friends, get more dates, get their points across better, or maybe even sell products if they happen to work in that kind of thing. For each of those situations, there are plenty of instances where it seemed that having a good sense of humour helped. People like to talk to someone funny at a party, a guy who can charm a woman with a few laughs does seem to have a better chance than someone who can't, and we all know advertisements that stuck with us because they were funny.

From an evolutionary point of view, though, humour never developed for anyone to deliberately manipulate it in order to manufacture friendships or bonds between people. Laughter is just a measurement for humans to evaluate the degree to which people are integrated into your social sphere. Out in the forests of our evolutionary past, things happened, people laughed, and then you knew they were with you, and similar to you. Only much later, when we had enough consciousness to be more deliberate in our outward expressions of internal feelings, did things like laughter, tears, and facial expressions become tools for use, both for social manipulations and for crafts like comedy and acting. The craft of comedy, then, is not well suited for making friends, as laughter is a guide for seeing how well you are getting along with people. It's not a source of friendship, or bonding of any kind, it's the result.

"But wait!" I hear you cry out, because you have an odd habit of speaking out loud to books you're reading. "I've seen people use humour as an ice breaker, with people they hardly know. How can you say we need to already be friends with people to make them laugh, even though I've seen people do it the other way around?" Good question. I'm glad you thought it so that I could assume you asked it. The answer comes from the difference between comedy and humour. Having got to this point in our evolution where we are self-aware enough to use humour as a tool of deliberate manipulation, the craft of comedy *can* be exploited to initiate the process of bonding between people. However, only in a limited fashion, and with diminishing returns.

A comedian is participating in a relationship that exists on a level of shared community. The world we live in, the society, the country, the city, the room, the crowd... there's a lot we have in common with the people around us even if we don't know them personally. A comedian takes that relationship to a community to its furthest theoretical extent. They aren't so interested in knowing each audience member on an individual basis. They are only looking for what they will *all* laugh at, as a group.

To a certain degree, that approach can be used as an icebreaker in a social context. Odds are that most parties I go to will involve a bunch of English speaking, middle aged and middle classed people, so I can probably draw upon a huge wealth of shared information that I can guess they are likely to have, even if I don't know for certain that any of them as individuals do. From a basis of that shared context, I can probably throw in a few ice-breaker jokes. But if you dropped me in, say, the middle of a village in a remote part of Pakistan, and asked me to make a class of school kids laugh, the challenge is harder. Not impossible, but one would have to reach down to some very broad concepts in the hope they have universal appeal for humans in general, which takes a lot of skill. Like *Monty Python* alumni Michael Palin, in one scene in his travel documentary series, *Himalaya*, where he entertains a group of Kalash school kids simply by chasing his shoe around.

As you can imagine, though, to carry on beyond a small amount of ice-breakers would be overly deliberate and puts one in a position of talking *at* people and not *to* them. Or, in other words, performing. That's not the goal in social circumstances. Successful social humour is not about our most broadly societal commonalities, as it is with performing in a comedy show, it's about our specific interpersonal commonalities, the ones we have between friends and the people we want to become our friends.

Ice-breakers are really a minor issue, then, in social humour. If you have a few witty things to throw around, whether they are old stand-by jokes that you read on the internet, or some amusing statements all of your own that you've done before, they will always have diminishing returns. After you get the conversation going, after a certain point, people want to *actually converse*, to have an interaction, not suddenly find themselves on the receiving end of someone's unilaterally imposed performance. The same is true in advertising situations as well. A good joke in an ad campaign can help you have a positive impression of a product, but there comes a line where, when actually purchasing it, you want to know what it's really about. So, if it's not crystal clear already, social humour and performance comedy are different. Performers are working on another level, appealing to groups of various scales that deal with similar circumstances. It's not going to help you to think like that if you want to get along with people you personally meet.

Let's not forget, too, that a lot of comedians are socially very awkward people. Those comedians have a way of observing and understanding themselves and the environment around them without it necessarily extending into an ability to socialize. It's entirely possible to have enough understanding of the world and yourself in it to make great material for comedy without being able to relate to the individuals around you.

Laughter should ideally flow naturally as a result of having strong social relationships. If it isn't, though, what can you do? Is there an understanding of humour that can help you enrich your relationships with friends and family? Yes, but to have a model of how to be more funny with people you feel you want to appeal to, you have to reconsider your motivations.

Nobody likes a person who imposes their wit on others. In other words, if you are attempting to be funny so that people will like you because of it, you're fighting a battle you ultimately can't win. Making laughter your first priority can cause people to see you as having a need to be the centre of attention, to be validated for how witty you are. That's not the kind of thing people look for in a friend, and your efforts to get laughter will become a wall between you and others. I don't say that lightly. I've been that guy.

How is it, then, that some people, are able to use humour to assist in their relationships? Not just with ice breakers when first meeting people, but to carry on beyond that, making humour and laughter a component of deepening friendships? Simple, really. They have their priorities straight. People first, humour second. A necessary first step in getting people to laugh in a social situation, where you don't already know everyone, is finding out about them. Then the jokes are not about you and how funny you are, but about them and what you understand about them. The key to having people be interested and amused by what you say is to be interested and amused by what they say. Genuinely interested, too. I don't propose this as scientific fact, but I believe that people are pretty good at detecting insincerity.

Getting to the point where you look beyond yourself and take an interest in others is the topic of many self help books, and the trade of many psychologists. It would be beyond the scope of this book to try and tell you how to address whatever personal demons keep you shy, or why it is you feel like the things that interest you are never interesting to others, or somehow you never feel as comfortable around others as you think successfully social people are, and so on.

If, however, you can get to know people a little, and then say something funny that derives from that unique interaction between you and them, a cycle can develop where a little laughter can put people at ease, which makes them let their guard down a little, which makes them more inclined to laugh, and the process repeats. At that point, you're well beyond ice breakers. As they let down their guard more, they share more about themselves, which becomes more potential material that we can share a point of view about. As people laugh together, each instance of laughing confirms among everyone in the group that they are getting along, and that they are sharing a point of view on a common experience.

Laughter is a useful, and powerful, tool in developing social bonds with people. The key is to recognize that laughter is your guide, not your goal.

😊 How To Learn

> *"You can't study comedy; it's within you. It's a personality. My humor is an attitude."*
>
> ~ Don Rickles

Emo Phillips, in an interview in a podcast, mentioned that one of the peculiarities of doing standup comedy is that a standup show begins with acts that are less talented and work their way up. You don't go to see the Chicago Symphony, and start the evening by watching the local high school jazz band struggle through some recently practised classics.

A rock band can get together in a garage and jam together, creating their own unique sound. There's an objective component to the ability to play their instruments and be harmonic that can feed their confidence about whether or not they are ready to perform in front of an audience. Whether or not people will like their songs is more easily identified as a separate matter of preference. Someone who wants to be a comedian, though, can't stand alone in a garage and perform jokes and develop much of a meaningful assessment of whether or not those jokes would even be considered jokes by anyone else.

A joke is the manifestation of a relationship between the comedian and the audience, and simply can not exist without both parties present. That's why comedians necessarily have to go through an open mike process, and why it's traditional for comedy shows to start out with comedians learning the craft. There's just no other way to do it. Just like you can't read a book about relationships and be a great spouse the next day, you have to get into a relationship in order to get good at it. As Jerry Seinfeld once remarked, "Being a good husband is like being a stand-up comic. You need 10 years before you can call yourself a beginner." No matter how much time, effort, skill, or any other resource a comedian spends alone working on material, there is no way of knowing whether or not it's any good until it's done in front of an audience.

Some comedians, particularly newer comedians, will try and find a way to practise their material before hitting the stage by sneaking jokes into conversations with friends. This is only really worth doing, though, if you can appreciate how your friend's relationship to you will impact how they take the joke. They have a whole body of context built out of their relationship with you that an audience does not have. Most jokes that will work with your friends won't work for that group of strangers. For that reason, I don't usually think its worth doing. For similar reasons, if you're looking to become a standup comedian, there are also drawbacks to inviting all your friends to come watch you perform. It's great to have their emotional support to help you get through the fears of performing, but they will still laugh at things that other people won't, and the more you intend to appeal to a broader group of people, the more you need to untether yourself from your friend's subjective approval of you.

All that said, writing material doesn't have to be an entirely solitary task, where you're completely in the dark about your potential until you get on stage. Before I got on stage to do standup for the first time, I turned to a friend who did some comedy, and asked her to listen to my material, to see if she thought it was funny. She understood comedy well enough to tell me in advance that she wasn't going to laugh, but that I shouldn't take that as any assessment of the potential of the jokes. Analyzing jokes to examine if they are funny creates a framework of pathways in your mind in which to receive the material. That deliberate process of connecting the material to the question "is this funny?" is a different pathway for receiving jokes, distinct from the kind of natural context an actual audience will have. Nonetheless, her advice at the time was helpful, and I still consult with other comedian friends when I think a joke of mine should be working, but I can't quite find the right way to make it happen.

Advice and tips are useful tools for honing a craft, and comedy is no different. Otherwise, there would be no point to this book. The key is knowing that the context of examining jokes is different from verifying them. Only an audience, the collective audience and not the individuals in it, the audience that doesn't care about why or how jokes work, and just want to laugh, can truly determine what is funny. You can get the best advice from the most established comedian, but you will never truly know a joke works until you've heard a real audience laugh.

☺ Finding Your Audience

"The bigger the audience, the better with comedy."
~ Jimmy Carr

Woody Allen once advised that when trying out new material, it's better to perform in front of the largest audience possible. To make that advice make sense, though, we have to put aside some of the practical considerations of trying to become a working comedian. If you're paid to perform at some huge venue, you're probably being paid to do the stuff that the booking agent knows is funny. You're not being paid to experiment with new material to see if it's any good. From a business perspective, it's generally good practise to work out the kinks with smaller venues where you're probably not getting paid much, if anything, and then when you get the chance to appear on a late night talk show and get seen by millions of viewers, then you present the best of what you have.

However, Woody Allen's advice is unrelated to the business aspect of making a comedy career. Becoming funny and making it big as a performer are different things. While developing an act, a comedian is likely to have experiences performing in all sorts of places of all sizes. Within that context, so long as we are only talking about the best way possible to discover what "works" as a joke, then Woody Allen's advice is to try the new material with the biggest possible audience that is available for your experimentation. In a sense, it's the natural logical extension of the point made in the last section, which is to move away from performing for individuals and head toward the larger context of communities. The larger the group you are performing for, the more indistinct the individual personalities within it become, and so the more you can be sure you are verifying your funniness against the group mind.

This runs against most comedians' natural instincts, which is to try new material with a smaller audience, and then enjoy the benefits of bringing the more trusted material in front of a large crowd. It's very satisfying to the ego to make a capacity crowd erupt with laughter, so who would want to waste that opportunity on new and untrusted material? Besides, for most comedians, the opportunity to perform to more than a handful of people who aren't their friends, or the other comedians performing that night, is a rare and cherished opportunity. The kind of opportunity you generally want to bring your A-game to. Fair enough. Probably the best compromise is already known to most comedians. Start off with some A-list material to let the audience know you're funny, work in some newer stuff in the middle to see how it goes, and then maybe end on some established material to finish strong. It's the kind of advice that truly advanced comedians scoff at because they find any formula demeaning to their craft. Rules are made to be broken. However, for comedians starting out, I think it's a reasonable enough guideline.

☺ Knowing What Works

> *"If you tell the truth about how you're feeling, it becomes funny."*
>
> ~ Larry David

One of my favourite comedy anecdotes has to do with Larry David. One time he came to a comedy club to perform a set. He got on stage, assessed the people in the room, and decided that his material was not going to work with that audience. He immediately got off stage without even trying a single joke.

If it's true that comedy ensues as a result of the right relationship with the audience, then maybe it's possible that someone with as much feel for comedy as Larry David can be attuned to the sensibilities of their audience to the degree that he can assess their potential to laugh at his material even before he starts. We'll never know for sure whether or not that audience would have laughed at Larry David's jokes enough to satisfy him, or at all. But, what it does tell us is that Larry David did not go in with the expectation that his material was what determined a successful performance. He assessed his relationship to the audience as being the main factor.

In most cases, though, simply dismissing the audience for not being able to relate is shirking responsibility. Larry David can back his decision to not perform that night with the fact that he has a wildly successful career, having made two critically acclaimed and hugely popular comedy television shows, on top of performing standup to large and appreciative crowds. Most of the time he proves that he can build the right relationship with his audience, which lends a lot of credibility to his decision that one time to not bother. For most comedians, backing down without at least trying will simply come across as cowardice.

The question is, *how hard* should you try? Where does one draw the line between jokes that are keepers and jokes that will never work? As a general rule, I've heard comedians say they try material out about three times, and if they don't get any laughs then it just isn't funny. However, if they get at least some response, even just a little, then it might be worth tweaking and editing until it becomes something better. That kind of advice exists elsewhere in many forms. For this book, what we want to know is *why* a joke told three times without getting a laugh isn't worth hanging on to.

Some relationships you have, like with your best friend, or spouse, or mother, are limited to single individuals. Other relationships, such as to people within your ethnicity or culture, are expansive on a scale of millions of people. Potential humour, being a product of relationship, is equally varied in how far its appeal will go. A joke that does not work three times in a row, if it is ever going to work for anyone, would appear to be more specific in its audience. Assuming that at one point you thought it was funny enough to consider using as comedy material, we know that at least one person had at one time the potential to laugh at it. That's a very specific audience. A joke that works with a wider audience is a more general meme that inspires the right activity among many individual minds.

Instead of thinking of jokes in simple terms of funny or not funny, it's more accurate to think of them as appealing to smaller or bigger audiences. On one end of the scale are the jokes that only made one person laugh when they thought of it themselves. At the other extreme end would be a joke that had the potential to make every human alive laugh. Which almost certainly does not exist, but it is still the theoretical maximum. Closer to the end where fewer people laugh are jokes among friends, and close to the other end where jokes have wide appeal is the domain of professional comedians who draw audiences of hundreds, or sell millions of books or downloads of their comedy specials. Somewhere in the middle are "local humour" jokes, the kind that you can tell in one city, but not take out on the road to another.

If a joke falls more toward the specific end of the scale, it's going to be harder to find an audience for it. Searching for the handful of other people who might laugh at it isn't really worth your time if you're going to try and perform comedy as anything more than a hobby. Especially when you also consider that you not only have to find those few people, but catch them at the right time, since their appreciation of your joke can vary depending on their particular state of mind at any one moment and how their feelings are influenced by the conditions around them. Really, a comedian in any media is looking for the jokes with the most universal appeal. Or, more accurately, the highest probability of effect within whatever social group they want to appeal to.

The three time test of jokes, then, is not really filtering out jokes on a binary scale of funny or not, it's just testing to see *how likely* it is you're going to come across an audience that likes it. You might try a joke more or less than three times, but it all has to do with how badly you want to find an audience for that joke. If you've told a joke ten times, and it isn't getting laughs, are you really determined to stick with it?

    In reality, it's unlikely that a comedian who is getting zero response on a joke is going to hang on to it for too long. From what I can see, a joke that gets stone cold silence is more likely to be abandoned too soon rather than too late. The real danger for comedians are the jokes that get lukewarm responses. There are as many different shades of laughter as there are types of joke. Some laughter can represent the audience being somewhat amused, and maybe for different reasons than the comedian intended. I've seen comedians hang on to jokes that get mediocre responses for years because it gets at least some laughter, so they think there's something to it. Good jokes are very precious, and I think most people who don't perform or write comedy don't realize how tough it is to develop material. Because it is so difficult, I don't hold it against any comedian who pads out their routine with some less than stellar material while they develop more solid stuff. The trick is that along the way they have to really listen to the audience to distinguish between what jokes are never going to work, which jokes are only going to go so far, and which jokes are gold. Ultimately, a successful comedian has to be able to discern not only which jokes don't work at all, but which jokes will only work so much. Letting go of underperforming jokes is a necessary step in replacing them with something better.

    In the end, whether you test a joke three times or more, or less, it's all about listening to the audience. As you get better as a comedian, one of the things you're doing is developing an ear for how the audience responds, because they are the ones who really know what's funny. Even better, you can get to the point of not only reading the audience response accurately, but also taking into account what kind of audience it is and honestly assessing how that influences their response. You could even get to the point where you can get on stage, and before even telling a joke, assess that your material and that audience are not going to mesh. Like Larry David. Just be prepared for the fact that no one is going to believe you didn't just chicken out if you don't have two hit comedy TV shows on your resume.

☺ It's All In The Delivery...?

> *"A comedian does funny things. A good comedian does things funny."*
>
> ~ Buster Keaton

In the same interview where Woody Allen spoke about the benefits of performing for a large audience, he also described how once he had some material that he thought was so great, so amazing, that he was sure he could just read it straight off the paper and get people to laugh. He tried it, and, predictably, it didn't work. His take away from that experience was that what the audience really wants is what he called "an intimacy" with the comedian. That seems to be in line with what this book says about comedy, but it would be an over simplification. A relationship with an audience is based on so many components. Think about all the relationships you have in your life. Your friends, your family, your loved ones. Are any of those relationships based on any one thing?

Putting together a comedy act, in whatever form, requires the combination of material, delivery, personality, and whatever else that works for the audience it hopes to reach. Putting all your eggs in one basket is a route to being a hack comedian, looking for the gimmicky shortcut to laughter, and not the real work.

The point really is that you are trying to build a connection to the audience, and in order for that connection to be novel to the audience, you have to do it your way. By virtue of the fact that all people are individuals, the intersection of all components of your act, including your material, your timing, your delivery, and everything else about how you perform, all coming together to varying degrees, will have a unique quality. That uniqueness is what creates the potential for inspiring new patterns of activity in the minds of the audience.

If you can encapsulate yourself into a performance that relates to the audience so that your uniqueness is still something people can relate to, you're in prime territory for comedy. How exactly to ride that razor's edge, though, is the art of it all. There is no one component of a performance that takes precedence over any other. You've got to assemble a style of comedy that represents you, and then give the audience all you've got. It all matters.

☺ Comedians Need You More Than You Know

> *"I love an audience. I work better with an audience. I am dead, in fact, without one."*
>
> ~ Lucille Ball

Clearly some people are better at being funny than others. Although no one can create a joke with certainty that it will get laughs, some people do manage to frequently create jokes with a higher probability of success than others. Comedians in all media, whether it's standups, authors, improvisers, or whatever else, are capable of becoming good enough at riding the waves of relationships that underlie comedy to base careers on, and even achieve international fame. How do they do it? If their source of talent can be identified, can you or I emulate it?

I believe that the answer is essentially yes, but with heavy qualification. To get at why I think the answer could be yes, I think it helps to know why it is that many people assume the answer would be no. It is a very common assumption that people who are funny have always been so. There are few, if any, stories of people who are known to be completely lacking in humour at one point in their lives, then going on to mastering the skills necessary to achieve some success in a medium with comedy. The assumption is always that they had some comedic ability, even if it was dormant and underused for a long time, and the change they made was to let go of some constraint so they could then follow their comedy dreams.

Humour and the ability to relate to other humans is something everyone has, so just like any human with legs could potentially learn to run a little faster, all humans could potentially participate in relationships in ways that facilitate more humour. That said, just because an ability to perform comedy isn't innate doesn't mean that there is nothing innate that can lead to comedy. Comedians might be the people who have a higher need to gain acceptance from the group. Or at least, a desire to receive the signal that says the group accepts them. That feature, that *desire* to hear laughter, that could be innate. Science already knows that people are born with predilections for certain stimuli that impact their overall happiness. Evidence indicates that alcoholics and drug users turn to substances because their brains have a certain need for particular sensations that satisfy them. Same for people with additions to sex, jogging, gambling, or whatever else. We all want stuff that keeps us happy and satisfied, and we seem to be born with different predispositions for which specific stimuli we prefer.

If some people are born with a desire for a particular stimulus, then one type of stimulus people could crave is laughter. Some people could crave it much more than others. Those people would gravitate toward behaviours that facilitate getting laughter, and do so from as soon as they are able, which can be from an early, even pre-verbal, age. When I was a toddler, I used to do impersonations of inanimate objects to make my mom laugh. For instance, I would throw a white blanket on the floor, crawl under it, and then twitch my arms and legs so that it looked like a frying egg. In my barely conscious, developing brain, getting her to laugh was what it was all about. If the desire to hear laughter motivates people to do something about it early on, any later success in that regard would very much appear to people observing the phenomenon to be an innate talent, something people were born with.

People who have always lived to hear others laugh have had a huge head start, honing the incredibly subtle crafts of timing and nuance that can make the difference between funny and unfunny. Not only would it be harder to catch up later in life simply in terms of time, but I would speculate that there is a difference between a person who feels an innate need to get laughs and one who objectively appreciates the social benefits from performance. People have a sense for sincerity, and I've seen many comedians die on stage because they wanted to be validated for being funny, as opposed to just wanting to make the audience laugh. A subtle, but crucial, difference that not everyone really appreciates.

It only seems like some people have innate talents that can't be emulated because those people have been working at getting their fix of laughter for longer than the people who are less capable. Realistically, not everyone can catch up with the people ahead on the curve, but anyone can potentially get at least a little funnier than where they are now.

☺ Stimulus And Response

> *"They used to laugh when I said I wanted to be a comedian. Well they're not laughing now!"*
>
> ~ Bob Monkhouse

One time I performed at a show, and it went really well. There was an audience of maybe sixty or seventy people, and the crowd laughed sincerely and loudly at every punchline I had hoped for them to laugh at. Or at least, it seemed that way, going by the best of my ability at the time to judge an audience response. On top of that, the other standups delivered great performances as well, so that the show overall appeared to me to be a roaring success. However, after the show and when some of the audience was shuffling out the door, I overheard two people speaking, and one of them casually remarked to the other, "Yeah, I thought the show was alright. A bit amateur, though."

What the hell...?

I was taken aback not only because I was bit deflated, but because I couldn't reconcile what I overheard with the laughter I experienced. If a comedian gets on stage and makes an audience laugh, haven't they completely succeeded in the task the audience expected of them? It seemed as weird as it seemed unfair that the audience could get all the laughs they hoped to get from a comedy show, yet still hope for something more, something additional and unquantifiable. What would separate a professional from an amateur if both get the same amount of laughs?

Most people would assume, as I did at that time, that laughter was the end result of all factors that go into assessing a comedy show. However, by understanding that laughter is, in some sense, an almost arbitrary response to certain stimulus in the mind alongside the thinking that people do as a matter of course, we can more easily appreciate that laughter is just one factor that contributes to an audience's impression of a show. It's a huge factor, maybe the most important, but it's still just one among many. When the audience is walking out the door, they will consider how much they laughed, but not stop there.

To be a successful performing comedian, you have to make an audience want to come back and see you again. They have to feel good about recommending you to their friends. To do that, it's not enough to just make them laugh. They also have to walk away comfortable with the thoughts and feelings you've dropped off in their brains. I'm not saying that you have to give them a heartwarming message or anything like that. As many successful comedians do, you can still go to extremes and edges of all manner, but the audience in the end has to have some kind of context where they feel okay about having those ideas in their head. Also that they respect the way in which you put them there. And that they believe you were in control of it. And everything else that can be a part of human perception. Simply put, it all matters. In retrospect, I think the reason I came across as amateurish in spite of also being funny was because didn't convey a level of mastery that made the audience believe I was fully in control of when I got laughs. Which, looking back, seems like a fair assessment.

Another common mistake newer comedians make is to mishandle jokes about the things I previously described as "uranium topics". Things like rape, paedophilia, racism, or anything that seeks to push the boundaries of acceptability. A lot of the time it fails and just results in an audience not laughing, but it can go another direction where an audience laughs, but don't hold a high opinion of the performer in the long term. In essence, what is happening is that a comedian can inspire the right synaptic activity in an area of the mind the audience doesn't really want to be in. It's a little like how you can tickle someone without them enjoying their laughter at all. It doesn't mean the audience has revealed any uncomfortable realities about their ethics or any hypocrisy in their identities. It only shows that the humour network is dispassionately listening for activity without any evaluation of the content. This makes it entirely possible for an audience to both laugh at something and also leave a show underwhelmed that the comedian merely used vulgarity as a crutch. It's unlikely that the audience will laugh with the same level of enthusiasm that they do with a comedian who is able to make uranium topics work in the long term impression as well, because when it's not working right, the audience's perceptions are simultaneous with the laughter, and both laughter and perception will feed back into each other, giving their laughter it's own particular quality. Nonetheless, they are still laughing, which might give some comedians a false sense of accomplishment in the short term.

This is true both on stage and off. You can act in a way that will get people to laugh, but not want to be your friend. If you want to be either a successful, funny friend or a successful comedian, laughter is your guide in the moment, but not the long term. For a comedian, this seems a little unfair. Laughter is hard enough to get, now on top of this, we're supposed to manage unmeasurable impressions beyond how funny we were? Impressions that are almost never going to be seen directly because most people will politely refrain from telling it to your face? Yep, that's the deal. No one ever said the route to comedic success was fair. The only way you're going to be able to tell if you've made that long term good impression is by much more subtle indicators, such as being asked to perform more and being given more opportunities. This level of feedback is much more subtle and organic, but it divides great comedians from mediocre ones. Which should make you appreciate the truly successful comedians all that much more, because they didn't just master being funny, they mastered long term impressions that don't have an immediate and automatic response.

☺ A Page More Blank

> *"I think the greatest thing that a performer can have if he's going to be successful as an entertainer is an empathy with the audience."*
>
> ~ *Johnny Carson*

The way to become a funnier comedian is already well known, and in practise anywhere that comedy happens. You get funnier by doing it. A statement that completely fails to satisfy anyone who is just starting out trying to become funnier. However, it's not that advice from experienced comedians, workshops, books on comedy, or any source isn't of value. Anything that provides an opportunity for you to think about how to become funnier, and provides any insights, can be helpful if it speaks to you in a way that inspires. The limiting factor is that the craft of comedy is built on developing a certain kind of relationship with an audience, and no one can become good at any relationship without being in it.

It's the same principle for any task that is built out of a relationship, like raising a child. You can read all the books on parenting you want, talk to all your friends who have kids, go to classes and seminars, and anything you gain from those sources is better than not having that information. No matter what, though, there are things you will *only* come to understand about being a parent when you have your own kids. It's not just that there are experiences so subtle and nuanced that they are hard to appreciate second hand. It's also that your child will be a different child, and you will be a different parent, and so yours will be a different relationship from all the relationships had by all the people you got advice from. Similarly, you are your own comedian, and the audience you have is your own. The one thing no advice, no workshop, no analysis of comedy, neither this book nor anything else, can ever give you in advance is a relationship to an audience. That has to be earned.

By seeing how the process of humour is more about how thoughts flow around in the mind rather than what those thoughts are about, we can come to truly appreciate how it is that you need to work with an audience to find out what it is that triggers their laughter. If humour were reliant on something contained within jokes, or humorous occurrences of any kind, then we could potentially isolate that magic and develop a predictive mechanism for telling future jokes. No such mechanism can ever exist, though, because the human mind can adapt and change. Predictability requires formality and structure, and the nature of the humour network is that it appreciates adaptability and change.

You can never be completely assured of where the audience mind is. Not only is any one human mind its own unique entity that you have no direct control over, as a performing comedian you are trying to relate to the collective result of the consensus of many unique minds, which results in the ethereal entity we know as an audience. Further, the audience, both as a whole and all the individual members in it, changes over time scales both small and large, so that what works now might not work again. Most of the time, for most potential humour, you had to have been there.

The one thing you have going for you is that despite all this wide open variance and possibility, humans will always try and come together. We want to get along because that's been the key to our success as a species. On a neuronal level, we're trying to build pathways in our brain that work well in conjunction with the pathways in other people's brains so that we can form communities that we can thrive in. The audience wants to laugh as much as you want to be funny.

Humour is a page far more blank than you might have ever imagined. It can never be constrained to any methodology, which is both its challenge and its charm.

Bibliography

"8 Simple Rules for Buying my Teenage Daughter." The Family Guy. Season 4 (episode 8). Fox. 10 July, 2005. Television.

Allen, Woody. Woody Allen on Comedy. Laugh.com: 2001. CD.

An Evening with Robin Williams. Dir. Don Mischer. Perf. Robin Williams. Vestron Video, 1982. Film.

"And the Weiner Is…" The Family Guy. Season 3 (episode 5). Fox. 8 August, 2001. Television.

Attardo, Salvatore, ed. Linguistic Theories of Humor. New York: Mouton de Gruyter, 1994. Print.

Azim, Eiman, et al. "Sex Differences in Brain Activation Elicited by Humor." PNAS 102.45 (2005): 16496 -16501. Web. 11 April, 2014.

Barry, Dave. "On Humor." News Writing Interviews. Annenburg Foundation, 2014. Web. 11 April, 2014.

Bell, Nancy D. "Responses to Failed Humor." Journal of Pragmatics 40.10 (2008): 1825-1836. Web. 11 April, 2014.

Bernard, Jill. "*Dramatic Improv: HARSH and Revolving Madness… Taking Improv Seriously in New York City and San Francisco.*" Revolving Madness, 2007. Web. 13 May, 2009.

"Best of the Rest Auditions." American Idol. Season 7 (episode 8). Fox. 6 February, 2008. Television.

Big Man Japan (Dai-Nipponjin). Dir. and Perf. Hitoshi Matsumoto. Yoshimoto Kōgyō, 2007. Film.

Borat: Cultural Learnings of America for Make Benefit Glorious Nation of Kazakhstan. Dir. Larry Charles. Perf. Sacha Baron Cohen. 20th Century Fox, 2006. Film.

Bressler, E. R., R. Martin, and S. Balshine. "Production and Appreciation of Humor as Sexually Selected Traits." Evolution and Human Behavior, 27 (2006): 121-130. Web. 11 April, 2014.

Carlsson K., et al. "Tickling Expectations: Neural Processing in Anticipation of a Sensory Stimulus." Journal of Cognitive Neuroscience 12.4

(2000): 691-703. Print.

Carson, Johnny. Johnny Carson On Comedy. Laugh.com, 2001. CD.

Carter, Judy. Stand Up Comedy: The Book. New York: Dell Publishing, 1989. Print.

"Cartoon Wars Part 1." South Park. Season 10 (episode 3). 5 April, 2006. Television.

"Cartoon Wars Part 11." South Park. Season 10 (episode 4). 12 April, 2006. Television.

Chomsky, Noam. "Chatting with Noam Chomsky." In Cohen, Henri, and Brigitte Stemmer, eds. Consciousness and Cognition: Fragments of Mind and Brain. London: Academic Press, 2007. Print.

Chris Rock: Bring The Pain. Dir. Keith Truesdell. Perf. Chris Rock. CR Enterprises, 1996. Film.

Cohen, Richard. "So Not Funny." The Washington Post 4 May 2006. Web. 9 Apr. 2014.

"Collection of September 11 related articles." The Onion. Web. 8 April, 2014.

Comedian. Dir. Christian Charles. Perf. Jerry Seinfeld. Bridgnorth Films, 2002. Film.

Cross, David. Shut Up, You fucking Baby! Sub Pop (USA) 2002. CD.

David, Larry. In Ricky Gervais Meets... Larry David. Dir. Niall Downing. Perf. Larry David, Rick Gervais, Andrew Sachs. Objective Productions: 2006. Film.

Dress To Kill. Dir. Lawrence Jordan. Perf. Eddie Izzard. Ella Communications Ltd., 1999. Film.

"Family Guy: Today Princeton, tomorrow the world." The Daily Princetonian. 5 February, 2004. Web. *8 April, 2009.*

Fried et al. "Electric Current Stimulates Laughter." Nature Magazine 391 (12 Feb., 1998), 350. Web. 13 April, 2014.

"Full Frontal Nudity." Monty Python's Flying Circus. Season 1 (episode 8). December 7, 1969. Television.

"Gender Differences are a Laughing Matter, Stanford Brain Study Shows." Stanford School of Medicine. 7 November, 2005. Web. 9 April, 2014.

Gervais, Matthew and David Sloan Wilson. "The Evolution and Functions of Laughter: A Synthetic Approach." The Quarterly Review of Biology 80.4 (December 2005): 395-424. Print.

Gilette, Amelia. "Eddie Izzard." A.V. Club, September 11, 2007. Web. 4 April, 2009.

Grenier, François, Igor Timofeev, and Mircea Steriade. "Focal Synchronization of Ripples (80-200 Hz) in Neocortex and Their Neuronal Correlates." The Journal of Neurophysiology 86. 4 (October 2001): 1884-1898. Print.

Gruner, Charles R. The Game of Humor: A Comprehensive Theory of Why We Laugh. Piscataway, NJ: Transaction Publishers, 2000. Print.

Goel, Vinod, and Raymond J. Dolan. "The Functional Anatomy of Humor: Segregating Cognitive and Affective Components." Nature Neuroscience 4 (2001): 237-238. Print.

Gottlieb, Sidney, ed. Hitchcock on Hitchcock: Selected Writings and Interviews. Berkeley: University of California Press, 1997. Print.

Hitchens, Christopher. "Why Women Aren't Funny." Vanity Fair, January 2007. Web. 1 April, 2009.

Hopper, Tristin. "Slurs Force Comic to Pay $15,000 for 'Tirade of Ugly Words' Against Lesbian Patron After Appeal Falls Flat." National Post. 23 June, 2013. Web. 9 April, 2014.

"How to Recognize Different Trees from Quite a Long Way Away." Monty Python's Flying Circus. Season 1 (episode 3), October 19, 1969. Television.

"Interview with Amy Sedaris." The Believer. March 2004. Web. 9 May, 2009.

Jerry Seinfeld: I'm Telling You for the Last Time. Dir. Matty Callner. Perf. Jerry Seinfeld. 1998. Film.

Johnstone, Keith. Impro: Improvisation and the Theatre. New York: Routlege, 1981. Print.

Johnstone, Keith. Impro for Storytellers. New York: Routlege, 1999. Print.

"Jon Stewart C-SPAN Interview." C-SPAN. October 25, 2004. Television.

Kreek, Mary Jeanne, et al. "Genetic Influences on Impulsivity, Risk Taking, Stress Responsivity and Vulnerability to Drug Abuse and Addiction." Nature Neuroscience 8 (2005) :1450 -1457. Print.

"Late Night with Conan O'Brien." Perf. Silverman, Sarah. 11 July, 2001. Television.

Leitch, Will. "Animal Magnetism." New York Magazine. 11 April, 2010. Web. 7 April, 2014.

Li, N. P., et al. "An Evolutionary Perspective on Humor: Sexual Selection or Interest Indication?" *Personality & Social Psychology Bulletin*, 35 (2009): 923-936. Print.

Linden, David J. The Accidental Mind. Cambridge, Mass: Harvard University Press, 2007. Print.

Marcus, Gary. Kluge: The Haphazard Construction of the Human Mind. Boston, MA: Houghton Mifflin, 2008. Print.

Martin, Demetri. Sound Of Young America. July 31, 2008. Radio.

Martin, Rod A. The Psychology of Humor: An Integrative Approach. Burlington, MA: Elsevier Academic Press, 2007. Print.

Maurer, Daphne, and Catherine J. Mondloch. "Neonatal Synesthesia: A Reevaluation." In L.C. Robertson and N. Sagiv, eds. Synesthesia: Perspectives from Cognitive Neuroscience. Oxford: University Press, 2005. Print.

Monty Python and the Holy Grail. Dir. Terry Gilliam, Terry Jones. Perf. Graham Chapman, John Cleese. EMI Films, 1975. Film.

Moran, Joseph M., et al. "Neural correlates of humor detection and appreciation." NeuroImage 21.3, March 2004: 1055-1060.

Murray, Michelle W. "Laughter is the 'Best Medicine' for Your Heart." University of Maryland Medical Center. 14 July, 2009. Web. 13 May, 2009.

"North by Northwest." Himalaya with Michael Palin. Episode 1, 2004. BBC. Television.

Olson, Ingrid R. and Christy Marshuetz. "Facial Attractiveness Is Appraised in a Glance." Emotion 5.4 (2005): 498-502 .

Panksepp, Jaak. "Beyond a Joke: From Animal Laughter to Human Joy?" Science 308 (2005): 62, 63.

"Patriot Games." The Family Guy. Season 4 (episode 20). Fox. 29 January, 2006. Television.

Paulos, John Allen. Mathematics and Humor: A Study of the Logic of

Humor. Chicago: University of Chicago Press, 1980.

Pav, Rich. "Kevin "Tokyo" Cooney." Herro Flom Japan, Podcasts & Videocasts from an American Salaryman. 27 Nov 2007. Web. 9 Mar. 2009 . Blog.

"Peter's Daughter." The Family Guy. Season 6 (episode 7). Fox. 25 November, 2007. Television.

Phillips, Emo. "Interview." Comedyspeak.com. Web. 30 September, 2008.

Pinker, Steven. The Stuff of Thought. New York: Viking, 2007.

Pinker, Steven. "The Colbert Report." Comedy Central. 7 February, 2007. Television.

Pinker, Steven. "What the F***." New Republic. October 8, 2007. Web. 7 April, 2014.

Provine, Robert R. Laughter: A Scientific Investigation. New York: Penguin Books, 2001.

Ramachandran, V.S., and Sandra Blakeslee. Phantoms in the Brain: Probing the Mysteries of the Human Mind. New York: William Morrow, 1998.

Reiss, A. L, et al. "Humor Modulates the Mesolimbic Reward Centers." Neuron 40.5 (2003): 1041-1048.

Robin Williams: At The Met. Dir. Bruce Gower. Perf. Robin Williams. Lions Gate, 1986. Film.

Scott, Sophie K. et al. "Positive Emotions Preferentially Engage an Auditory–Motor 'Mirror' System." The Journal of Neuroscience. 26.50 (2006):13067-13075.

Seinfeld, Jerry. Comedians on Comedy: Jerry Seinfeld. Laugh.com: 2001. CD.

Seinfeld, Jerry. Seinlanguage. New York: Bantam, 1994.

"Seinfeld Star Richards under Fire for Racial Outburst." Reuters, November 20, 2006. Web. 9 April, 2014.

"Smart and Smarter." The Simpsons. Season 15 (episode 13). 22 February, 2004. Television.

Stableford, Dylan. "Remembering The Onion's 9/11 issue: 'Everyone thought this would be our last issue in print'." Yahoo News. August 25, 2011. Web. 8 April, 2014.

Stanley, Alessandra. "Who Says Women Aren't Funny?" Vanity Fair. April 2008. Web. 1 April, 2009.

Star Trek II: The Wrath of Khan. Dir. Nicholas Meyer. Perf. William Shatner, Leonard Nimoy. Paramount Pictures, 1982.

The Aristocrats. Dir. Paul Provenza. Perf. George Carlin, Chris Rock. Mighty Cheese Productions, 2005. Film.

The Knickerbocker, or New-York Monthly Magazine. XXIX. John Allen, 1847.

"The Pilot Episode." Committed. Season 1 (episode 1). NBC. 4 January, 2005. Television.

The Secret Policeman's Biggest Ball. Dir. Mike Holgate. Perf. John Bird. October, 1989. Film.

"The Year-End Mega-Matrix." New York Magazine. December 10, 2006. Web. 09 Apr. 2014.

Vanity Fair. "Christopher Hitchens: Why Women Still Aren't Funny." YouTube. 3 Mar., 2008. Web. 1 April, 2009.

"Whither Canada?" Monty Python's Flying Circus. Season 1 (episode 1). October 5, 1969. Television.

Index

Abbot and Costello, 137
acting, 23, 178, 180, 203
adaptive bias, 91
advertising, 204
alcohol, 181
alien, 56, 57, 60, 61
Allen, Gracie, 156
Allen, Steve, 126
Allen, Woody, 149, 150, 209, 210, 216
American Idol, 65
Anderson, Willy, 105
Aoki, Guy, 94-96
Aristotle, 26
axon, 39, 115
axon terminals, 39, 40, 41
backwards rationalizing, 92
Ball, Lucille, 217
banter, 163, 186
Barnum, PT, 124
Barry, Dave, 136
Bell, Nancy, 63
benign violation, 26, 27, 79
bias, 66, 110, 146, 151, 201
Bierce, Ambrose, 42
Big Man Japan, Dai Nipponjin, 167
bigotry, 94, 95, 163, 167
bilateral posterior temporal lobes, 21
Borat (movie), 103
Borge, Victor, 16
Boston Improvisation group, 177

Brand, Jo, 183
Brandes, Jurgen, 142
Bugs Bunny, 137
bullying, 24
Burns, George, 3, 5, 156
Bush, George W., 164
Caesar, Sid, 70
Carlin, George, 96
Carr, Jimmy, 209
Carroll, Lewis, 8
Carson, Johnny, 224
cats, 9, 21, 71
Chaplin, Charlie, 117, 138
Chaucer, 137
chicken crossing the road joke, 131
chimpanzees, 21, 27, 28
Chomsky, Noam, 8, 134
CK, Louis, 132-134
Clay, Andrew Dice, 94, 95
Cleese, John, 62
Cohen, Sacha Baron, 103
cohesion, 59, 61, 64, 65, 68, 74, 150
Colbert Report, The (TV show), 33
Colbert, Stephen, 33, 164-166
collective agreement, 177
collective minds, 145
collective relevance, 175
Comedian (Jerry Seinfield documentary), 118
comedy ray, 195, 196

Committed (TV show), 142, 145
common evolutionary ancestor, 70
commonality, 59, 60, 66, 70, 71, 74, 76, 82, 148, 188, 200
community relevance, 198
computers, 6, 7, 159-161
concept, 8, 37, 42, 43, 45, 49, 51, 56, 64, 68, 79, 95, 101, 107, 108, 119, 148, 157, 160, 166, 178, 188, 195, 196, 204
consensus, 9, 15, 21, 45, 46, 64, 67, 143, 144, 167, 175-177, 225
context, 4, 16-18, 44, 51, 61, 62, 64, 65, 67, 78, 80-82, 93, 98, 100, 106-108, 110, 118-120, 124, 128, 130, 131, 133, 134, 137, 144, 150, 155, 162, 163, 167, 168, 170, 175, 176, 178, 179, 186, 187, 198-204, 208-210, 221
Cross, David, 149
cuisine analogy, 109
culture, 9, 26, 27, 68, 87, 108, 110, 148, 150, 152, 160, 167, 168
Cummings, Ray, 115
Dai Nipponjin, Big Man Japan, 167
dam, 46-49
dam analogy, 46
David, Larry, 149, 212, 215
delivery, 10, 15, 27, 77, 216
dendrites, 39, 181
Descartes, Rene, 169
desire, 142, 149, 163, 188, 218, 219
Dietrich, Marlene, 136
dogs, 21, 71
Dostoyevsky, Fyodor, 80
drag, 153
drugs, 182, 183, 195
Earle, Guy, 161
Eastman, Max, 93
electrochemical, 42, 44
electrochemical signal, 40, 66, 115
ethics, 81, 94, 96, 97, 222
evolution, 59, 61, 64, 69, 71
evolutionary development, 67
evolutionary time scales, 68
existentialist thinking, 79
expectation, 10, 12, 65, 81, 87, 88, 121, 131, 146, 147, 149, 152, 154, 156, 163, 182, 188, 200, 202, 212
eye colour, 56, 59
factual reality, 176
false positives, 69
Family Guy, The, 80-82
Ferguson, Craig, 174
Fey, Tina, 154, 193
flow of activity, 44, 45, 50, 66, 79, 92, 93, 97, 100, 106, 114, 116, 119, 120, 137, 168, 188, 195
Frankl, Viktor E., 99
Fried, Itzhak, 37
friend funny, 19, 83, 155, 201
Friends (TV show), 110, 149
Fry, Stephen, 46
funeral, 9, 76, 105-107

gap analogy, 31
gap model, 50
gender equality, 151
ghetto, 146
Giraudoux, Jean, 22
Gravas, Latka, 157
grey matter, 152
group dynamics, 144
groups, 57
guideline, 58, 59, 211
Hamada, Masatoshi, 187
Handler, Chelsea, 180
Hardy, Thomas, 129
hate speech, 161
Hebb, Donald, 40
heckler, 163, 182-187
Hesiod, 114
Hicks, Bill, 161
Himalaya, 204
hindsight bias, 90, 95, 102
Hitchens, Christopher, 151
horror movies, 17, 65
horse, funny, 37
humorous occurrences, 11
humour centre, 21, 34, 37
humour network, 37, 38, 46-48, 50, 52, 58, 61-63, 65, 70, 72-75, 79, 82, 87, 92, 93, 95, 97, 98, 105, 106, 116, 122, 157, 162, 168, 170, 182, 183, 186, 194, 195, 222, 225
humour sensation, 67
hypocrisy, 222
ice breaker, 203
Impro For Storytellers, 177
incongruity resolution, 26
inhibition, 174, 181, 182, 198
insincerity, 167, 206

intimacy, 216
intuitions, 79, 144
ippatsu gagu, one hit gag, 169
Irony, 164-170
Japanese, 78, 81, 109, 110, 148, 166, 167, 169, 187
Jewish comedians, 149
Jewish comedy, 148
Jewish culture, 149, 150
Johnstone, Keith, 176, 177
Joyce, James, 88
Kaufman, Andy, 157, 158
Keaton, Buster, 137, 138, 216
Kids In The Hall, 153
Kierkegaard, Soren, 77
Klein, Robert, 150
Knickerbocker Magazine, 132
Knight, Heather, 159
language, 8, 21, 26-28, 40, 51, 72, 73, 78, 90, 97, 107, 109, 167, 168
laugh button, 36
laugh centre, 33, 34, 48, 50, 194
Laugh Lab, The, 14, 25
Laurel and Hardy, 137
local humour, 214
long term strengthening, 115
love songs, 17, 65
Luther, Martin, 102
M, 115
macrotiming, 115, 126, 128
manipulation, laughter as a tool for, 23
Martin, Demetri, 75

Marx, Groucho, 65
mating dance theory, 151
Matsumoto, Hitoshi, 167, 187
medial ventral prefrontal cortex, 21
Meiwes, Armin, 142
meme, 12, 134, 143, 213
mesotiming, 115, 117-119
metacontext, 79, 81, 103
metacontextual thinking, 79
microtiming, 115-119
Milligan, Spike, 14
Modest Proposal, A, 136
Monkhouse, Bob, 220
Monty Python, 82, 129, 130, 153, 204
Moran, Joseph, 21
motor cortex, 20, 33, 34, 38, 105, 194
Muntz, Nelson, 22
neonatal synethsesia, 43
neuronal network, 43, 44, 94, 102, 119, 198
neurons, 38-42, 47, 48, 51, 58, 66, 78, 101, 102, 106, 115, 130, 181, 195
neuroplasticity, 42
neurosurgery, 36
New York magazine, 164
Nietzsche, Friedrich, 56
No Apologies, 94
nucleus accumbens, 21
offensiveness, 89, 94
Office, The (TV show), 145
one hit gag, 170
Onion, The, 127, 165
Palin, Michael, 204
Pardy, Lorna, 161

Parker, Trey, 80
People's Daily Online, China, 165
perception, 10, 27, 43, 45, 51, 72, 78, 102, 133, 143, 144, 156, 167, 176, 189, 221, 222
phenotype, 60, 146
Philips, Emo, 197
Pilkington, Karl, 33
Pinker, Steven, 33
pins, 11-13, 15, 34-36, 201
plosives, 25
Pokhis, Yakov "Smirnoff", 107
politically correct, 89
pop culture, 80, 81, 147
porn, 17, 65, 88
potential humour, 8, 11, 13-16, 18, 34, 36, 45, 46, 49-51, 78-80, 88, 89, 91, 92, 108, 115-119, 130, 131, 133, 162, 182, 213, 225
predictability, 225
primal limbic, 98
Provine, Robert, 3, 18
Pryor, Richard, 145, 146
puns, 26
purring, 21
Quinn, Colin, 118, 187
rats, 21, 70
relationships, 9, 13, 17-19, 21, 31, 38, 66, 67, 69, 89, 104, 110, 131, 137, 156, 157, 159, 160, 170, 186, 188, 193, 197, 198, 200, 201, 203, 205, 207, 208, 212, 213, 216-218, 224, 225
research, 3, 13, 15, 21, 36,

37, 43, 60, 63, 76, 97, 154, 159
Richards, Michael, 161
Rickles, Don, 207
Rivers, Joan, 150, 151
Rock, Chris, 31, 98
roses, 16, 27, 134
Rudolph, Maya, 154
sakura, 174
Sando, Katsura, 170
sarcasm, 165-167
satire, 94, 96, 164-166
Saturday Night Live, 147, 154
scanner, 13, 20, 22, 32-34, 36, 37, 40, 98
Schrödinger's punchline, 12
Sedaris, Amy, 154
Sedaris, David, 136
Seinfeld (TV show), 149
Seinfeld, Jerry, 31, 32, 45, 50, 76, 118, 149, 150, 208
sequence, 18, 118, 133
sequential experience, 118
sequential relationship, 195
Seuss, Dr., 38
sexism, 151
Shakespeare, William, 20, 136, 137
Shaolin Kung Fu Masters, 13
shared patterns, 101, 145
Shaw, George Bernard, 105
shock humour, 102
Silverman, Sarah, 93-96
Silverstein, Andrew Dice Clay, 94
similarity, 58, 59, 61
Simon, Neil, 25

Simpson, Lisa, 22
Simpsons, The (TV show), 22
sincerity, 22, 24, 76, 166, 219
slapstick, 137
Smith, Mark, 164
Smoove, Jay, 65, 67
soccer players, 23
social humour, 204
somatosensory, 20
South Park, 80
sphere of expectation, 176-179
Spitting Image (British puppet show), 166
spontaneous, 33, 186
spotlight bias, 147
stage funny, 19, 83, 155, 201
Stanley, Alessandra, 154
Star Trek, 56
status, 23, 25, 27, 51, 59, 79, 143
stereotypes, 145, 146, 148-150, 152, 163
stimulus, 11-13, 15, 18, 23, 36, 201, 218, 221
Stimulus and response, 7, 12, 13, 18, 220
Stone, Matt, 80
Strangers With Candy (TV show), 154
study, Joseph Moran, 21
study, research, 3
study, Stanford, 152
study, The Laugh Lab, 14
subarachnoid hemorrhage, 105
Sumitani, Masaki, 170

super hero comics, 70
surprise, 25, 27, 51
survival, 57
swear words, 96
Swift, Jonathan, 136, 164
sympathy, 23
synapses, 39, 41, 44, 62, 97, 115, 130, 133, 199
synaptic connections, 42-44, 101, 114, 116, 168
taboo, 97-102, 107
Takeshi's Castle (Japanese game show), 110
talent, 144, 154, 155, 202, 217, 219
television analogy, 27
the tilt, 179
tickling, 20, 70-73, 194, 195
timing, 18, 26, 114, 115, 117, 118, 128, 134, 153, 183, 184, 216, 219
Tomlin, Lily, 142
Tonari No Seinfeld (My Neighbour Seinfeld), 149
too close, 31, 46, 50, 51, 62-64, 79, 81, 95, 97, 122, 126, 128, 129, 145, 149, 157, 179
too far, 31, 46, 50, 51, 62, 64, 66, 79, 82, 122, 145, 157, 158, 178
tragedy, 127
truthfulness, 89
Tsu, Lao, 35
Turing, Alan, 159
Twain, Mark, 44, 136

two cows joke, 25
two ducks joke, 25
two hunters joke, 14
umor, 6
Un, Kim Jong, 165
uranium, 100, 221
Vanity Fair, 151, 154
vaticinium ex eventu - hindsight bias, 90
visual gags, 28
Voltaire, 50
Washington Post, 164
Washington State University, 63
Watterson, Bill, 120
weak connectivity, 51
weak synaptic connections, 46, 58, 81, 102, 107, 108, 196
White House Correspondents' Association dinner, 164
white matter, 152
White, E.B., 6
Wiig, Kristen, 154
Wilde, Oscar, 201
Williams, Robin, 124-126
Wiseman, Richard, 14, 25
Wittgenstein, Ludwig, 36, 134
working blue, 96
Wright, Steven, 12
Yoshio, Kojima, 170
zaniness, 137, 169, 170

www.ingramcontent.com/pod-product-compliance
Lightning Source LLC
Chambersburg PA
CBHW071712170526
45165CB00005B/1989